David Henry Browne

Nickel-Steel: A Synopsis of Experiment and Opinion

David Henry Browne

Nickel-Steel: A Synopsis of Experiment and Opinion

ISBN/EAN: 9783337138950

Printed in Europe, USA, Canada, Australia, Japan

Cover: Foto ©berggeist007 / pixelio.de

More available books at **www.hansebooks.com**

Nickel-Steel; A Synopsis of Experiment and Opinion.

BY DAVID H. BROWNE, CLEVELAND, OHIO.

(California Meeting, September, 1899.)

INTRODUCTION.

THE trite saying that man is a tool-using animal might now-a-days be amended by saying that man is a tool-choosing animal. The chipped flint, at first all-sufficient, gave way to hammered bronze, and that, in turn, to wrought-iron and steel. As our knowledge of metals increases, we learn to specialize; we temper and anneal; we alloy and intermix; and we fashion for ourselves a metal suitable to our necessities. The knowledge of the properties and uses of the various metals and their alloys has become the capital of every constructive engineer, and the currency of every intelligent brass-founder and blacksmith. From year to year, the specifications of the designer become more rigid, and the steel-maker is forced to study more closely the qualities obtainable by changing the treatment, or altering the composition, of his metal.

Of making alloys there is no end. We have heard much, during the last decade, of manganese-steel and chrome-steel, of aluminum and tungsten, of titanium- and silicon- and copper-steels, and of every conceivable union into which steel could be coerced or cajoled. Without denying to each of these alloys novel and valuable properties, it must be admitted that, of all alloys, the one which shows the greatest range of adaptability, and has met the largest measure of popular approval, is nickel-steel. Introduced to public notice by Mr. Riley, in 1889, its field of usefulness has been rapidly enlarged, until, at the present day, the use of nickel-steel is as thoroughly recognized as that of carbon- or manganese-steel.

Concerning the properties of nickel-steel numerous articles have appeared in the technical journals. For the man in a hurry, who wishes to look up the properties and limitations of nickel-steel, these articles might almost as well be in the lost library of Alexandria. The purpose of this paper is to present in systematic form a summary of the best that has been written upon this subject. Some hitherto unpublished data have been secured; the opinions of manufacturers and users have been collected; and the various uses to which nickel-steel is applied have been investigated and recorded.

My thanks are due to Mr. E. F. Wood, of The Carnegie Steel Co.; Mr. H. F. J. Porter, of The Bethlehem Steel Co.; Mr. G. H. Chase, of The Midvale Steel Co.; Mr. Henry Souther, of The Pope Manufacturing Co.; Mr. H. J. Williams, of the Damascus Tool Co.; and Mr. H. K. Landis, of New York, for procuring new data, and for kindly aid in obtaining comparative tests. It is hoped that the material thus collated may prove interesting and useful to all who are in any way concerned with the manufacture and use of steel.

Physical Qualities of Nickel.

The three metals, iron, manganese and nickel, form a group having remarkably similar physical and chemical qualities. In specific weight, atomic weight, atomic volume, melting-point and color, these metals have a close family resemblance. While an examination of any one of these reveals no remarkable physical superiority over the others, the fact remains that each seems to exercise, in an alloy, a beneficial effect upon the other two. The addition of small amounts of these metals to one another produces alloys much superior in physical qualities to the pure metals. Manganese increases the strength of pure nickel; and both manganese and nickel greatly increase the strength of pure iron.

The most careful investigations yet made upon the alloys of nickel and iron are those of the committee appointed in 1891 by the Berlin "Society for the Encouragement of Technical Industry." The reports of this committee, compiled by Professors Wedding and Rudeloff, and published from time to time in the *Verhandlungen des Vereins zur Beförderung des Gewerbfleisses*, will form, when completed, a systematic record of the properties of the alloys of nickel and iron. As these papers have not yet been translated into English, and as the original reports are not easily accessible in the United States, I have given, in the following pages, very full excerpts from them.

For these investigations, Professor Rudeloff procured samples of the best commercial nickel and of iron almost free from carbon and other impurities. These metals were cast into bars, rolled into rods and straps, and tested in the raw condition, and also after hammering and annealing. A series of alloys of nickel with pure iron was then made; and a complete series of tests was instituted, to show the effect of nickel in amounts varying from zero to 98 per cent. The commercially pure nickel test-bars, after melting and casting, were found by analysis to contain:

	Per cent.
Nickel,	98.13
Cobalt,	1.15
Iron,	.43
Silicon,	.08
Magnesium,	.11

The physical properties of this pure nickel in the bar as cast and after rolling into flat bars were as follows:

TABLE I.—*Physical Properties of Nickel Bars.*

Commercial Nickel.	Limit of Proportionality. Lbs. per Sq. In.	Elastic Limit. Lbs. per Sq. In.	Ultimate Strength. Lbs. per Sq. In.	Prop'rtion Elastic to Ultimate. Per cent.	Modulus of Elasticity.	Elongation. Per cent.
Cast bars	5,119	12,557	40,669	38	23,989,140	18.2
Rolled { Raw	9,243	21,045	72,522	29	29,506,500	43.9
Rolled { Annealed	17,064	18,059	72,806	25	26,870,800	48.6
Rolled { Quenched		16,921	71,860	24		45.0

For purposes of comparison, the physical properties of the pure iron (containing 0.07 per cent. carbon) used for the preparation of the nickel-iron alloys, are here given:

TABLE II.—*Physical Properties of Iron Bars.*

Pure Iron.	Limit of Proportionality. Lbs. per Sq. In.	Elastic Limit. Lbs. per Sq. In.	Ultimate Strength. Lbs. per Sq. In.	Prop'rtion Elastic to Ultimate. Per cent.	Modulus of Elasticity.	Elongation. Per cent.
Cast bars	8,532	20,761	46,072	45	32,030,550	36.8
Rolled { Raw	20,761	31,568	50,196	63	28,795,500	42.3
Rolled { Annealed	21,756	32,848	44,793	73	29,151,000	49.8
Rolled { Quenched	9,100	48,774	69,962	70	31,568,400	27.8

Hadfield, in his recent paper on nickel-iron alloys,* gives the following tests of nickel:

TABLE. III.—*Cast and Forged Nickel* (98.8 *Per Cent. Nickel*).

Test-Bar.	Elastic Limit. Lbs. per Sq. In.	Tensile Strength.	Elong. in 2 in. Per cent.	Contr. of Area.	Fracture.
Cast, unannealed............	24,640	36,400	4.5	9.76	Granular.
Forged, unannealed........	31,360	72,128	45.5	57.04	Fibrous, silky.
Forged, annealed............	15,680	70,000	54.	52.5	Fibrous, silky.

The physical tests of metallic nickel furnish no explanation of its remarkable effects when alloyed with iron. Nickel has a specific weight of 8.3 to 8.9; its atomic weight is 58.6, and its atomic volume 6.60. It melts at about the same temperature as steel, its melting-point depending on the amount of carbon present. The specific heat of nickel is about 0.108 to 0.110. Nickel is slightly stronger than pure iron, somewhat harder, and much less easily oxidized by moisture and steam. It can be heated to redness without perceptible oxidation, is slowly soluble in hydrochloric and sulphuric acid, and is readily soluble in dilute, but remains passive in concentrated, nitric acid.

Nickel can retain carbon, both as graphite and as amorphous carbon; it can absorb silicon, and will alloy in any proportion with iron. It absorbs oxygen when melted, and requires the addition, when molten, of about 0.1 per cent. of magnesium, in order to form a sound casting.

Alloyed with iron and steel, nickel produces a remarkable series of metals, the physical qualities of which will be taken up in systematic order.

II.—PHYSICAL QUALITIES OF NICKEL-STEEL.

Elasticity.

Nickel-steel is most emphatically distinguished from simple steel by its high elastic strength. While this factor is, to a certain extent, shown by the test for elastic limit, it must

---

* "Alloys of Iron and Nickel," *Trans. Inst. Civil Engrs.*, London, vol. cxxxviii., Part iv., 1899.

not be confounded therewith. The strength of a man may be measured in three ways. We may first determine the actual load which his body, rigidly suspended, could sustain without actual rupture. This load would correspond to the ultimate tensile strength of steel. Secondly, we may determine the actual load which, with the utmost bodily exertion, he can sustain for a few seconds. This would correspond to the elastic limit of steel as usually tested by "drop of beam." Thirdly, and with entire justice, we may determine the amount of work which he can daily and regularly perform without undue fatigue and with no injurious strain to any bodily organ. This is properly compared to the elastic or live strength of steel, as measured by the French "limit of proportionality," or by the resistance of the bar to numerous and rapidly alternating strains. It is a well-known fact that rupture may be produced by numerous applications of a load below the elastic limit of the metal; and such an occurrence in the case of shafts and axles is vaguely ascribed to the "fatigue of metals" in service. This term conveys a larger scientific truth than is generally appreciated; and in some testing-works it is customary to measure more or less empirically the endurance of metals with regard to what may be termed molecular fatigue. If steel were a homogeneous metal, every applied load, less than the elastic limit, should produce a deformation proportionate to the load; and upon the removal of the load the test-piece should regain its original dimensions. If, for example, on a steel rod under tension, a load of 1000 pounds produces a stretch of one thousandth of an inch, then 2000 pounds should produce double this stretch; and 4000 pounds should cause four times the deformation produced by 1000. If we find that by successive increments of 1000-pound loads we get regularly increasing stretch, proportioned to the load, up to 10,000 pounds, but that a load of 11,000 pounds produces twelve-thousandths of an inch stretch, it is evident that, at 10,000 pounds' load, the ratio between strain and stretch changes. This point is called by the French and German investigators the "limit of proportionality." It is far below the elastic limit of the test-piece. In the case we have outlined, the load might be increased to 20,000 pounds before a permanent set would be produced; but it is very evident that the elastic strength of a metal is more accurately measured by the limit of proportionality, where a molecular fatigue begins, than by the commonly known "elastic limit," where an actual molecular distortion takes place.

The addition of nickel to iron and steel increases the ultimate strength, and, to a higher degree, the elastic limit; while its greatest influence is shown on the limit of proportionality. Some exceedingly interesting experiments conducted by Prof. Rudeloff,* in Berlin, on the alloys of varying amounts of nickel with almost chemically pure iron, bring out this feature very strikingly.

TABLE IV.—*Elastic Strength of Iron-Nickel Alloys.*

Iron containing Nickel. Per cent.	Limit of proportionality. Pounds per square inch.	Iron taken as standard.
0	8,532	100
0.5	8,816	103
1	10,238	120
2	14,504	170
3	22,894	268
4	23,605	275
5	27,871	326
8	32,421	380
15	22,752	266

An alloy of 8 per cent. of nickel with pure iron has 3.8 times the elastic strength of pure iron. With higher percentage of nickel the limit of proportionality decreases, the greatest effect being shown at 8 per cent. The effect of nickel on the elasticity is not changed by the presence of carbon; and for this reason nickel-steel shows, as Mr. Davenport says, "that combination of high elastic strength and toughness" which distinguishes it from simple steel.†

At the Bethlehem Steel Works, in Bethlehem, Pa., and at the Pope Tube Works, in Hartford, Conn., it is customary to measure the elastic strength of metals by their resistance to numerous and rapidly alternating flexures. The apparatus consists of an ordinary lathe, with a ball-and-socket chuck, into which one end of the test-piece is fastened, while the other end is supported by a collar in the tail-stock. A roller-bearing

---

* *Verein zur Beförderung des Gewerbfleisses*, Berlin, February, 1896.
† "Steel for Marine Forgings," *Cassier's Magazine*, August, 1897.

ring encircles the test-piece at its center, and from this depends an adjustable weight.

"Assume," says Mr. Souther, "that in all cases the specimens are annealed, and by 'annealed' I mean treated at the best annealing-temperature for each grade of steel. Take a 0.10 per cent. carbon-steel as a basis, loaded say to two-thirds its elastic limit, maximum fiber-stress; assuming that this, under these circumstances, will withstand 100,000 revolutions, then a 0.25 carbon-steel will stand somewhere in the neighborhood of 200,000; a 0.50 carbon = about 400,000; and nickel-steel (0.25 carbon, 5 nickel) in the neighborhood of 1,000,000 revolutions."*

In these tests the live strength or working-life of the metal is shown, and a very close estimate can be made of the value of a metal under normal conditions of work.

Elastic Limit and Ultimate Strength.

The elastic limit and tensile strength of steel are so much influenced by the percentage of carbon present, and by the mechanical or heat-treatment to which the metal has been subjected, that it is of the utmost importance to secure for comparison test-pieces in which all modifying conditions, except the percentage of nickel, are identical.

The most complete tests made up to the present time are those of Rudeloff, on almost carbon-free nickel-alloys; and of Hadfield on nickel-steels of very low and almost uniform carbon-contents.

TABLE V.—*Strength of Nickel-Iron Alloys.*—(Rudeloff.)

Iron Containing Nickel. Per cent.	Elastic Limit.		Tensile Strength.	
	Lbs. per Sq. In.	Compared with Iron as Standard.	Lbs. per Sq. In.	Compared with Iron as Standard.
0	20,761	100	46,072	100
1	23,605	113	47,921	104
2	28,724	137	52,614	114
3	34,128	184	57,733	125
4	38,251	184	57,733	125
5	46,215	222	63,421	137
6	62,852	303	79,916	173

Mr. Hadfield's experiments on steels containing very low

---

* Mr. Henry Souther, Engineer Pope Tube Co., private communication.

carbon (0.15 to 0.20), with different amounts of nickel, show the following results :*

TABLE VI.—*Annealed Test-Bars (Cast).*

Nickel. Per cent.	Elastic Limit. Lbs. per Sq. In.	Tensile Strength. Lbs. per Sq. In.	Elongation. Per cent.	Contraction. Per cent.
0.27	44,800	62,720	37	52
0.51	47,040	60,480	41	63
0.95	44,800	60,480	41	63
1.92	49,280	69,440	36	53
3.82	56,000	73,920	35	55
5.81	62,720	82,880	33	51
7.65	67,200	100,800	26	41
9.51	71,680	125,440	2	2
11.39	100,800	199,360	12	26
15.43	....	152,320	1	1
19.64	100,800	194,880	5	4
24.51	56,000	174,720	14	8
29.07	35,840	82,880	48	51
49.65	33,600	80,640	49	53

According to the experiments of Rudeloff, the addition (up to 5 per cent.) of nickel to pure iron produces for each 1 per cent. of nickel added a gain of 5090 pounds in the elastic limit, and a gain of 3469 pounds in the ultimate strength. Hammered test-bars of the same material, show a gain of 5403 pounds elastic, and 5175 pounds ultimate strength for each 1 per cent. of nickel added. Fig. 1 represents the approximate curve of increase in elastic limit due to 1 per cent. of nickel added to steel of different carbon-contents.

If nickel be added to the soft steels, it raises the elastic limit and the ultimate strength in much the same proportion as in pure iron. Carbon does not therefore prevent the influence of nickel; but, as will be explained later, nickel has a very marked effect upon the carbon present.

The proportion of nickel commonly used in soft steels for armor and for engine-forgings is from 3 to 3.5 per cent. This percentage of nickel, alloyed with an open-hearth steel of 0.25 per cent. carbon, produces a metal equal in elastic limit and tensile strength to a similar open-hearth steel of 0.45 carbon without nickel. In other words, 3 per cent. of nickel, added to a soft steel, has the same effect on the strength as the intro-

---

* "Alloys of Iron and Nickel," *Trans. Inst. Civil Engrs.*, London, vol. cxxxviii., Part iv., 1889.

duction of 0.20 per cent. of carbon would have, while the ductility remains at the figure usually found in the lower-carbon steel. Protective deck-steel of 0.45 carbon and no nickel will have a tensile strength of 90,000 pounds, with 15 per cent. elongation in four diameters; while a 0.25-carbon steel, with

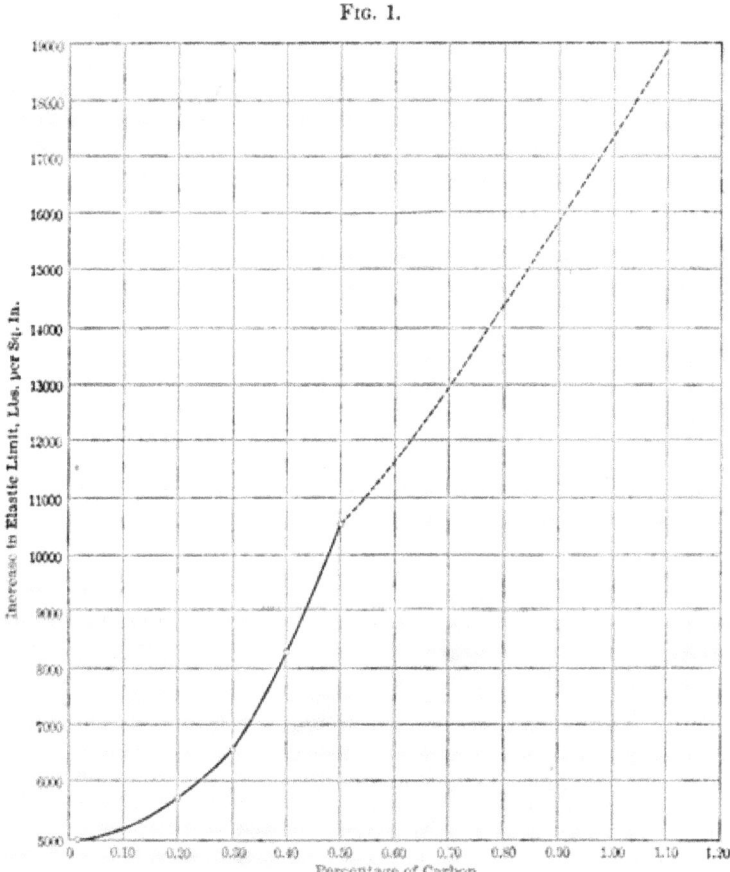

Fig. 1.

APPROXIMATE CURVE OF INCREASE IN ELASTIC LIMIT DUE TO ONE PER CENT OF NICKEL

3 per cent. of nickel, will have the same tensile strength of 90,000 pounds, with 20 per cent. of elongation, in four diameters.*

On all the low-carbon steels the addition of nickel seems to produce remarkably uniform results. A flange-steel (0.10 car-

---

* Mr. E. F. Wood, Carnegie Steel Works. Private communication.

bon and no nickel), rolled into sheets, gave 35,243 pounds elastic and 54,450 pounds ultimate strength. The addition of 2.7 per cent. of nickel brought these figures up to 47,080 and 65,750 pounds respectively.* This shows a gain of 4384 pounds on elastic limit and 3814 pounds on ultimate strength for each 1 per cent. of nickel added.

Steel tubing (of 0.25 carbon and no nickel) has very closely 40,000 pounds elastic limit and 70,000 pounds ultimate strength. The addition of 5 per cent. of nickel raises the elastic limit to 60,000 and the ultimate to 95,000 pounds per square inch— the gain for each 1 per cent. of nickel added being 4000 pounds on the elastic and 5000 pounds on the ultimate strength.†

Rolled steel, of 0.14 carbon, tested by Mr. Allan, of Glasgow, showed the following comparison with nickel-steel:

Carbon. Per cent.	Nickel. Per cent.	Elastic Limit. Lbs. per Sq. In.	Ultimate Strength. Lbs. per Sq. In.
0.14	none	27,552	63,168
0.14	3	43,008	73,696

The gain was 5485 pounds on the elastic limit and 3509 pounds on the ultimate strength for each 1 per cent. of nickel added.‡

On homogeneous irons and low-carbon steels, not subjected to heat-treatment, it appears that each 1 per cent. of nickel added, up to 5 per cent., causes an increase of about 5000 pounds elastic limit and 4000 pounds tensile strength.

With higher percentages of carbon, the influence of nickel is divided between its action *per se*, producing a gain in toughness and strength, and its action on the carbon present, modifying to a certain extent the original influence of nickel.

High percentages of nickel tend to throw any carbon in the alloy into the graphite state. A 30 per cent. nickel-steel, containing 1 per cent. of carbon, will have 0.80 carbon in the shape of graphite.§ A 25 per cent. nickel-steel retains carbon

---

* Canadian Copper Co.'s experiments.

† Mr. Henry Souther, Engr. of Tests, Pope Mfg. Co. Private communication.

‡ Beardmore, "Nickel-Steel, an Improved Material," etc.

§ Rudeloff, *Verein zur Beförderung des Gewerbfleisses*, Heft I., p. 85.

in a state in which the metal is not influenced by tempering.* Carbon existing in pure nickel is almost entirely in the graphitic condition; and high percentages of nickel tend to throw the carbon into this state. With lower percentages of nickel, the effect is to change part of the carbon from cement-carbon to hardening-carbon, thereby rendering the metal exceedingly sensitive to heat-treatment.†

The Bethlehem Steel Company gives the following table of comparative tests of nickel steels and simple steels on oil-tempered, annealed forgings:‡

TABLE VII.—*Comparative Tests of Nickel-Steel and Simple Steel Forgings.*

Composition of Forgings.	Elastic Limit. Lbs. per Sq. In.	Tensile Strength. Lbs. per Sq. In.	Elongation. Per cent.	Contraction of Area. Per cent.
No nickel, 0.20 carbon	28,000	55,000	34	60
3.5 per cent. nickel, 0.20 carbon	48,000	85,000	26	55
Influence of 3.5 per cent. nickel	+20,000	+30,000	−8	−5
" 1 per cent. nickel	+5,714	+8,571		
No nickel, 0.30 carbon	37,000	75,000	30	50
3.5 per cent. nickel, 0.30 carbon	60,000	95,000	22	48
Influence of 3.5 per cent. nickel	+23,000	+20,000	−8	−2
" 1 per cent. nickel	+6,571	+5,714		
No nickel, 0.40 carbon	43,000	85,000	25	45
3.5 per cent. nickel, 0.40 carbon	72,000	110,000	18	40
Influence of 3.5 per cent. nickel	+29,000	+25,000	−7	−5
" 1 per cent. nickel	+8,285	+7,142		
No nickel, 0.50 carbon	48,000	95,000	21	40
3.5 per cent. nickel, 0.50 carbon	85,000	125,000	13	32
Influence of 3.5 per cent. nickel	+37,000	+30,000	−7	−8
" 1 per cent. nickel	+10,570	+8,571		

This table is exceedingly interesting, as showing that the effect of nickel on the elastic limit increases as the carbon increases.

---

* Cholat and Harmet, *Oesterr. Zeitsch. für Berg-und Hüttenwesen*, 1894, p. 543.

† Mr. Chase, Midvale Steel Co. Private communication.

‡ Mr. R. G. Davenport, Mr. H. F. J. Porter and Mr. Hart. Private communication.

In the 0.20-carbon steel, the gain on elastic limit due to 1 per cent. of nickel is 5714 pounds; while in the 0.50-carbon steel, the gain on elastic limit, due to 1 per cent. of nickel, is 10,570 pounds.

Mr. McDonald, of the Crescent Steel Company, of Pittsburgh, reports a test of nickel in forged steel of 1.10 per cent. carbon, as follows:

Composition.	Elastic Limit.	Tensile Strength.
No nickel, 1.10 carbon,	57,700	112,452
1.65 per cent. nickel, 1.10 carbon,	88,980	146,250
Gain due to 1.65 per cent. of nickel,	31,280	33,798
" to 1 " "	18,957	20,484

The influence of nickel on both elastic and ultimate strength seems, therefore, to increase with the percentage of carbon; or, in other words, high-carbon steels show a greater gain from the addition of nickel than low-carbon steels. While the above data are too few to base accurate comparisons upon, still the reports made by the Bethlehem Iron Company coincide very well with the results given by other steel works; and as a basis for future comparison the gains due to 1 per cent. of nickel have been plotted in the curve shown in Fig. 4.

Cholat and Harmet found that in pure iron the greatest strength was obtained by the addition of 15 per cent. of nickel. Taking this 15 per cent. alloy, and adding to it varying amounts of carbon, they found the greatest strength was produced by 0.30 carbon, at which percentage a tensile strength of 213,400 pounds was obtained. The same alloy (0.30 carbon, 15 nickel), oil-tempered, showed an elastic limit of 166,300, and an ultimate strength of 277,290 pounds per square inch.*

Experimenting on steel containing from .15 to .20 carbon, Mr. Hadfield finds the greatest strength is produced by the addition of 12 to 15 per cent. of nickel. Cast test-bars, not annealed, containing 11.39 per cent. nickel, showed a tensile strength of 210,560 pounds per square inch, the same figure being also given by a steel containing 15.48 per cent. nickel.†

---

* Cholat and Harmet, *Oesterr. Zeitsch. für Berg- und Hüttenwesen*, 1894, p. 534. Abridged in *Scientific American*, Jan. 8, 1897.

† Hadfield, "Alloys of Iron and Nickel," *Inst. Civil Engrs.*, 1899.

Elongation and Contraction of Area.

Nickel-steels show a slight decrease in elongation and contraction of area when compared with simple steels of the same percentage of carbon, but an increase in these measurements, when compared with simple steels of the same tensile strength or elastic limit.

A 0.25-carbon structural steel, annealed, tested by J. G. Eaton,* showed·

Carbon. Per cent.	Nickel. Per cent.	Elastic Limit. Lbs. per Sq. In.	Ultimate Strength. Lbs. per Sq. In.	Elongation. Per cent.	Reduction of Area. Per cent.
0.25	none	36,420	62,410	26.25	58.5
0.25	3.3	59,500	84,640	24.50	54.0

Comparing steels of equal carbon, it appears that nickel produces a slight decrease in elongation and contraction of area.

If steels of equal tensile strength be compared, nickel-steel will show greater ductility than carbon-steel:†

Carbon. Per cent.	Nickel. Per cent.	Elastic Limit. Lbs. per Sq. In.	Ultimate Strength. Lbs. per Sq. In.	Elongation. Per cent.	Reduction of Area. Per cent.
0.20	3.5	45,000	85,000	26	50
0.40	none	43,000	85,000	25	40

On protective deck-plates, simple steel of 0.45 carbon will show 90,000 pounds tensile strength and 15 per cent. elongation in a 2-inch piece; while a 0.20-carbon steel, with 3 per cent. nickel, having the same tensile strength of 90,000 pounds, will give 20 per cent. of elongation in the same length.

"With a given tensile strength," says Mr. R. W. Davenport,‡ "nickel increases somewhat the elongation, and to a greater extent the contraction, of area."

It may therefore be taken as a rule that, as compared with simple steels of the same tensile strength, a 3 per cent. nickel-steel will have from 20 to 30 per cent. greater elongation; while, as compared with simple steels of the same carbon-percentage, the nickel-steel will have about 40 per cent. greater tensile strength, with practically the same elongation and re-

---

* Commander J. G. Eaton, "The Advantage of Nickel-Steels over Carbon-Steels."

† Mr. E. F. Wood, Carnegie Steel Co.   ‡ *Cassier's Magazine*, Aug., 1897.

duction of area. This relation of ductility to tensile strength holds good only for the usual nickel-steels, containing up to 5 per cent. of nickel.

In the alloys of nickel with pure iron, the experiments of Wedding and Rudeloff, and those of Cholat and Harmet, Guillaume and Moulan, have shown a different relation.

The figures obtained by Rudeloff for cast bars of alloys of nickel, with almost pure iron, show the following relations between elastic limit, ultimate strength and ductility:

TABLE VIII.—*Strength and Ductility of Nickel-Iron Alloys.*

Nickel. Per cent.	Elastic Limit. Lbs. per Sq. In.	Ultimate Strength. Lbs. per Sq. In.	Elongation. Per cent.
0	20,761	46,072	29.7
1	23,605	47,921	26.4
2	28,724	52,614	22.7
3	34,128	57,733	20.1
4	38,251	57,733	17.6
5	46,215	63,421	10.8
8	62,852	79,916	9.6
16	....	58,302	0.6
30	....	14,077	2.2
60	17,775	53,751	36.1
93	15,357	47,210	19.0
98	12,940	43,371	17.1

From this it will be seen that in alloys of nickel with iron containing little or no carbon the strength of the alloy increases with each rise in nickel up to 8 per cent., while the ductility decreases up to 16 per cent. Beyond this point, and up to 60 per cent., the increase of nickel causes an increase both in ductility and strength.

At about 25 per cent. of nickel it seems that a peculiar molecular change is produced. A wire with 24 per cent. of nickel is as soft and ductile as copper,* and a 27 per cent. nickel-steel shows 47 per cent. of elongation, with a tensile strength of 100,000 pounds per square inch.† Riley states‡ that 25 per cent. nickel-steel, containing 0.27 carbon, gave an elongation of 29 per cent., with a tensile strength of 102,600 pounds per square inch; while nickel-steel, with 25 per cent.

---
* Guillaume, *Comptes Rendus*, March, 1898.
† J. C. Danziger, *Detroit Eng. Soc.*, June 19, 1896.
‡ J. Riley, *Jour. Iron and Steel Inst.*, May, 1889.

of nickel and 0.82 of carbon, gave 40 per cent. of elongation, with a tensile strength of 94,300 pounds. In this case it will be noticed that a large increase in carbon gave a large increase in ductility, with a slight drop in tensile strength. Cholat and Harmet state that in 25 per cent. nickel-steels, elongation and reduction of area rise with increase in the carbon-content, and the tensile strength remains high.*

These tests show that the steels containing high percentages of nickel are entirely different in physical properties from the low-nickel steels commonly used in mechanical engineering.

Proportion of Elastic Limit to Ultimate Strength.

In low-carbon simple steels the elastic limit is about 50 per cent. of the ultimate strength. The addition of nickel to soft steel and homogeneous iron increases the proportion of elastic limit to ultimate strength, as is shown by the following tables, taken from Rudeloff's experiments:†

TABLE IX.—*Proportion of Elastic to Ultimate Strength in Iron-Nickel Alloys.*

Nickel. Per cent.	Proportion Elastic to Ultimate. Per cent.
0,	45
1,	49
2,	55
3,	59
4,	66
5,	73
8,	79

M. Moulan,‡ of the Société Cockerill, at Seraing, Belgium, in experimenting on an alloy of pure iron, with 7.5 per cent. of nickel, found that in soft iron the ratio of elastic limit to ultimate strength is 55 per cent; while in 0.55-carbon steel the ratio is 51 per cent.; and in the alloy of homogeneous iron with 7.5 per cent of nickel, the ratio is 91 per cent. of the ultimate. The same samples having been oil-tempered and annealed at 500° C., the ratios were found to be 65 for iron, 74 for steel, and 96 for the ferro-nickel alloy. This high propor-

---
* *Oesterr. Zeitsch. für Berg- und Hüttenwesen*, 1894, p. 543.
† *Verein zur Beförderung des Gewerbfleisses*, Berlin, 1896.
‡ *Industries and Iron*, Nov. 2, 1894.

tion of elastic limit to ultimate strength had been attained without loss of ductility, since in the oil-tempered and annealed sample the elongation of the ferro-nickel alloy was 12.2 per cent.

"The ratio of elastic limit to tensile strength (in steel) can be increased by tempering (i.e., hardening by sudden cooling), with subsequent annealing, and also by the introduction into the metal of a proper amount of certain unusual ingredients, notably nickel."*

"Nickel," says Mr. Colby,† "increases the ratio between the elastic limit and tensile strength, and adds to the ductility of the steel." Tests made by Mr. Colby show that on forgings of mild steel, annealed, the ratio between elastic limit and ultimate strength is 46.4; in medium hard steel, annealed, 46.2; and in medium hard nickel-steel, annealed, 58.7 per cent.

Mr. Allan, Lloyd's Surveyor at Parkhead, Glasgow, finds the ratios as follows:

	Proportion of Elastic Limit to Ultimate Strength. Per cent.
Forged nickel steel,	64.0
" simple "	54.4
Rolled nickel "	58.3
" simple "	43.6

Mr. Beardmore states that the ratio of elastic limit to tensile strength is from 63 to 74 per cent. in 3 per cent. nickel-steel forgings.

The influence of nickel upon the strength of steel is most apparent in the elastic strength or working-capacity of the metal. If simple carbon-steel of a given tensile strength be compared with nickel-steel of the same tensile strength, it will be found that the elastic limit of the nickel-steel is from 10 to 20 per cent. greater than that of the carbon-steel. This effect of nickel upon the elastic limit and the limit of proportionality accounts for the increased working-capacity of nickel-steel and its resistance to molecular fatigue.

Influence of Mechanical Work on Nickel-Iron Alloys.

A very thorough investigation of the effect of forging, and flat- and round-rolling, on the physical properties of nickel-

---

* R. W. Davenport, *Cassier's Mag.*, Aug., 1897.
† A. L. Colby, *Trans. Engrs. Club*, Phila., 1897, vol. xiii.

steel has been made at the Royal Mechanico-Technical Experimental Station, at Charlottenburg, Prussia. The average results obtained are given in Table X., in which the qualities of the various alloys of nickel and iron, after rolling, are compared with their qualities in the cast bars. This table should be studied in connection with Table VIII.:

TABLE X.—*The Effect of Hot-Rolling on Nickel-Iron Alloys*—The Values in the Cast Bar taken as* 100.

Nickel. Per cent.	Elastic Limit after Rolling.	Ultimate Strength after Rolling.	Elongation after Rolling.
0	160	102	136
1	156	102	132
2	140	104	155
3	132	101	161
4	118	107	186
5	113	107	301
8	97	94	318
16	......	315	1211
60	197	164	92
95	159	172	213
98	104	175	243

From this table it will be seen that the elastic limit of the low-grade nickel-steels is not as much increased by rolling as is that of wrought-iron. The figures of increase in ultimate strength are, for the low percentages of nickel, about the same as for pure iron. The ductility of nickel-steel, however, is increased by mechanical treatment to a much greater extent than that of iron. The 16 per cent. nickel-iron alloy shows a remarkable increase in ultimate strength and ductility after rolling. The actual figures obtained were as follows:

TABLE XI.—*Tests of* 16 *Per Cent. Nickel-Iron Alloy, as Cast and as Rolled.*

	Ultimate Strength. Lbs. per Sq. Inch.	Elongation. Per cent.
Cast,	58,302	0.6
Rolled and annealed,	183,580	11.0

This extraordinary increase in the strength of a 16 per cent. nickel-iron alloy by working is corroborated by the declaration of Mr. H. J. Williams,† that an 18 per cent. nickel-steel re-

---

* *Verein zur Beförderung des Gewerbfleisses*, vi. and vii., 1898.

† H. J. Williams, Damascus Steel Co., Pittsburgh, Pa. Private communication.

TABLE XII.—*The Effect of Heat-Treatment on Nickel-Iron Alloys.*—(Tables from Wedding and Rudeloff.)

Elastic Limit in Pounds per Square Inch.

Percentage of Nickel.	None.	1.	2.	3.	4.	5.	8.	16.	30.	60.	93.	96.
Raw, cast.............	20,761	23,605	28,724	34,128	38,251	48,215	62,852		17,750	17,750	15,357	15,357
Hammered............	27,871	39,105	40,669	41,650	46,783	54,889	75,223			14,788		26,449
Hammered and annealed.....	34,554	40,953	41,664	46,783	45,788	54,889	70,246			27,871		8,816
Hammered and quenched.....	46,926	50,765	56,026	87,168	105,651	137,931	161,254*			33,843		12,229

Tensile Strength in Pounds per Square Inch.

Raw, cast.............	46,072	47,920	52,614	57,733	57,733	63,421	79,916	58,302	14,077	53,750	47,210	40,669
Hammered............	45,930	52,398	54,016	62,141	63,950	71,811	87,737			18,059		68,398
Hammered and annealed.....	44,508	51,476	52,614	58,444	62,994	69,678	84,324			50,196		63,421
Hammered and quenched.....	65,698	76,788	79,347	112,622	136,512	183,764	161,254*			60,130		53,803

* Elastic limit and tensile strength here appear to be the same.

ceives its highest obtainable tensile strength and elasticity by hammering on the anvil from a red heat until cold.

Influence of Heat-Treatment on Nickel-Steels.

On heating simple carbon-steels to redness, and plunging them into oil, the elastic limit and tensile strength of the metal are raised, the effect being greater on the high-carbon than on the low-carbon steels. Prof. Howe, in his *Metallurgy of Steel*, gives tables showing that the tensile strength is increased by quenching from 23 per cent. in 0.10-carbon steel to 77 per cent. in 0.50-carbon steel; while, in the same samples, the elastic limit was raised by this treatment 83 per cent. in the soft and 161 per cent. in the hard steel.

In the absence of carbon, nickel itself can render iron very sensitive to heat-treatment; and in the presence of carbon the sensitiveness to heat-treatment which carbon induces is augmented by the influence of nickel. The effect of nickel alone is shown in Table XII., in which tensile tests of iron almost free from carbon, but containing nickel, are compared in four conditions: (a) cast bars; (b) hammered bars; as (1) hammered, (2) annealed and (3) quenched.

From this table it will be seen that the effects of annealing and quenching are as follows: On alloys containing 8 per cent. nickel, and less, annealing lowers to a slight extent the tensile strength and elastic limit. By quenching, the elastic limit of nickel is uniformly raised.

If the strength of the hammered and annealed bar be taken as 100, the gain in strength produced by quenching is as follows:

TABLE XIII.—*Increase of Strength in Iron-Nickel Alloys by Quenching.*

(In Percentages upon the Figures for the Hammered and Annealed Bar before Quenching.)

Nickel. Per cent.	Increase in Elastic Limit. Per cent.	Increase in Ultimate Strength. Per cent.
0,	36	48
1,	24	50
2,	34	51
3,	86	93
4,	131	117
5,	147	164
8,	130	91

These figures, from Wedding and Rudeloff,* are corroborated by the experiments of M. Moulan, of the Société Cockerill.† Comparing "homogeneous iron," containing almost no carbon, with the same metal after alloying with 7.5 per cent. of nickel, he found the gain in tensile strength and elastic limit produced by heat-treatment to be as follows:

TABLE XIV.—*Increase of Strength Produced in Iron, and in a Certain Iron-Nickel Alloy, by Quenching.*

(In Percentages upon the Results given by the Corresponding Metal before Quenching.)

	In Oil.		In Water.	
	Elastic Limit. Per cent.	Ultimate Strength. Per cent.	Elastic Limit. Per cent.	Ultimate Strength. Per cent.
Homogeneous iron,	50	15	31	4
Alloy containing 7.5 per cent. nickel,	96	84	116	132

Nickel of itself can therefore confer upon its alloys with iron the property of taking increased strength by quenching. The higher nickel-alloys (over 16 per cent.) are not affected by heat-treatment. When, in connection with the properties due to nickel alone, the effect of the carbon present is considered, the extreme sensitiveness to heat-treatment of the ordinary grades of nickel-steels may be understood and appreciated.

Annealing-Temperature of Nickel-Steels.

If steel be allowed to cool gradually from the solidification-point, it will be noticed that the fall of temperature is at first regular; at a certain point the cooling ceases, and the metal for a time either remains at a constant heat or even shows an increase in temperature; below this point the metal cools regularly as before. This point where cooling is interrupted marks the temperature at which steel changes from the amorphous and plastic condition, in which it can be forged and welded, into the crystalline structure which it retains when cold. It marks the temperature at which the best results can be produced by quenching, and it marks also the temperature to be attained (and very slightly exceeded) in annealing. The larger

---

* Rudeloff, *Verein zur Beförd. des Gewerbfl.*, vii. and viii., 1898.

† "Le Ferro-Nickel," *Rev. Univ. des Mines*, vol. xxvii., 1894, p. 142. The table here given is not directly quoted, but formed by calculation from M. Moulan's results.

the proportion of carbon present, the lower is the recalescent point, and the narrower are the ranges of temperature within which the steel can be safely worked. A 1.50-carbon steel can be refined (i.e., brought to its maximum strength and hardness) by quenching from a dark orange red. A steel of 1.00-carbon can safely be heated to a little brighter redness; and a 0.50-carbon steel will bear a bright orange heat, which would burn and destroy the texture of a harder steel.

"The presence of nickel causes steel to crystallize at a much lower temperature than it would otherwise."* This statement, by Mr. Ellis, managing director of John Brown & Co., of Sheffield, is explained by an examination of the recalescent point of nickel-steel. The determination of this point has been made by Messrs. Henry Souther and F. S. Flavel for the Pope Manufacturing Company, of Hartford, and is shown, in comparison with steels of 0.25 and 0.50 per cent. carbon, in the curve exhibited in Fig. 2. It appears that the recalescent point of 0.25-carbon steel is a little over 1600° F., while that of 0.50-carbon steel is between 1350° and 1400°, and that of 5 per cent. nickel-steel, with 0.25 carbon, is about 1080° F. The recalescent point of pure nickel is 1112° F.† This furnishes an explanation of the super-sensitiveness of nickel-steel to heat-treatment. The proper annealing-temperature for the simple carbon steel is, according to Mr. Souther, a full red, while for nickel-steel the heat should not be over a cherry-red.

The Welding of Nickel-Steels.

Nickel has a great affinity for oxygen, which renders the welding of high-nickel steels a matter of difficulty. Among steel-makers the term "welding" is often used to express the hammer-treatment of a cast ingot at a bright red heat to insure the absence of blow-holes and internal flaws or cracks. Among the steel-users, welding means exclusively the union by hammering or pressure of two pieces of steel at the proper temperature. All nickel-steels may be consolidated and forged into homogeneous masses under the hammer or hydraulic press. In other words, the presence of large amounts of nickel does not render

---

* "Recent Experiments in Armor," *Inst. Naval Architects*, March, 1894.
† "Nickel Extraction by the Mond Process," W. C. Roberts-Austin, *Inst. Civil Engrs.*, London, Nov., 1898.

steel red-short. For blacksmiths' work, however, an increase in the amount of carbon, or an increase in the amount of nickel, renders welding difficult. A perfectly welded union should be as strong at the points of contact as at any other part of the welded piece; and hence a job of welding which a blacksmith might deem satisfactory might not show good results in a testing-machine.

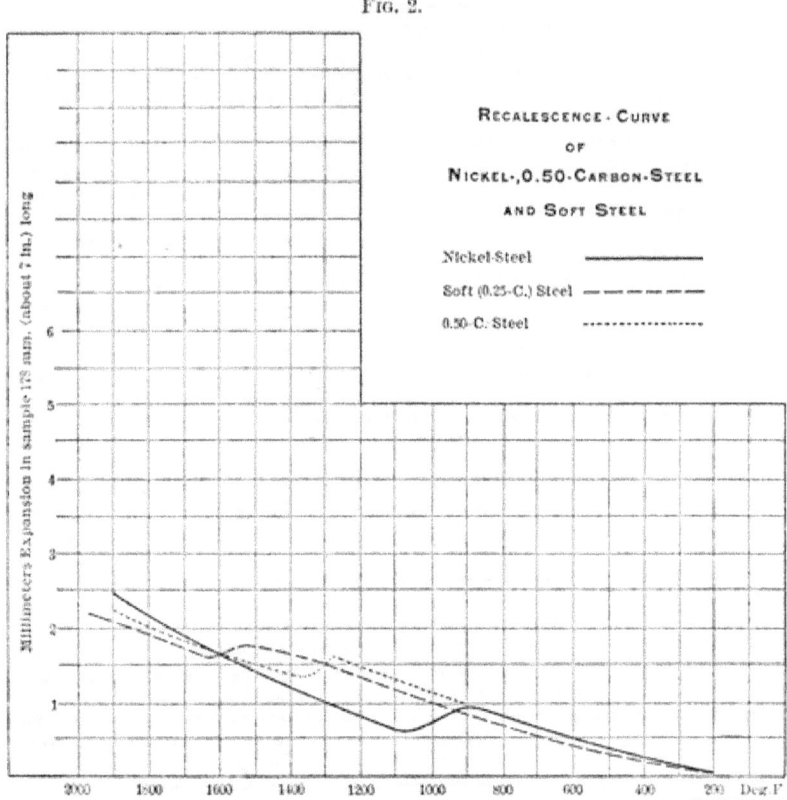

FIG. 2.

The writer has had the following tests made by a skillful, practical blacksmith:

(1) Steel, 2.05 Ni., .22 C., cut like soft steel; welded perfectly; no sign of weld showing; bent twice at right angles at the weld when hot; the weld did not open, nor was any crack noted; bent at right angles when cold; failed to show any crack at the weld.

(2) Samples containing 3.25 per cent. nickel and 0.16 carbon worked exactly like No. 1, under the same tests; no crack was developed, and the metal welded perfectly.

(3) Samples containing 3.40 per cent. nickel and 0.31 carbon cut a trifle harder, also hammered like a harder steel; welded perfectly; bent hot and cold, like No. 1, and showed no crack. The weld could not be seen.

(4) Samples containing 2.62 per cent. nickel and 0.19 carbon worked exactly like Nos. 1 and 2. The tests did not show any weakness at the weld.

(5) Samples containing 3.20 per cent. nickel and 0.54 carbon worked a little harder, but gave a perfect solid weld. There were no cracks on bending, whether hot or cold.

(6) Samples containing 3.10 per cent. nickel and 0.96 carbon worked harder (i.e., like a tool-steel); welded perfectly, and showed no cracks on bending hot or cold.

(7) Samples containing 4.95 per cent. nickel and 0.51 carbon worked like No. 5, but not as hard as No. 6. The weld was good, and no cracks developed on bending.

These tests are more practical than scientific. The writer has also seen an 18 per cent. nickel-steel split-welded to soft carbon-steel: the weld, ground down on an emery-wheel, showed apparently perfect union. Tests of the same metal for strength and elongation in welded and unwelded samples are likely to be deceptive, since the presence of nickel renders the carbon very sensitive to the heat-treatment received in welding, and, therefore, unless the welded and normal bars are subjected to the same annealing before testing, the results are discordant.

Mr. Hadfield, in his recent paper on the "Alloys of Nickel and Iron" (cited above), expresses the opinion that it is not possible to produce a perfect weld in nickel-iron alloys. Admitting that, with very high percentages of nickel, welding is impossible, I must, however, maintain that Mr. Hadfield's conclusions are not entirely corroborated by the experience of others. Some very interesting tests have been made by the Canadian Copper Co. on the loss of strength in welded specimens of nickel-steel. Two test-pieces of each grade were prepared; one was tested in its normal condition, while the other piece was cut in two, electrically welded, and then tested. The results are shown in table on page 24.

Tests made at the Royal Prussian Testing Institute show that the average tensile strength of welded bars of medium steel (no nickel) was 58 per cent. of that of the original bars before welding. In softer steels (no nickel) the average strength was 71 per cent., while in puddled iron the average was 81 per cent.

Beardmore* gives carefully-conducted welding-tests on 0.26-

---

* "Nickel-Steel as an Improved Material for Boiler-Plates, etc."

TABLE XV.—*Tests of Electrically-Welded Nickel-Steel.*

No.	Carbon. Per cent.	Nickel. Per cent.		Tensile Strength. Lbs. per Sq. In.	Elongation. Per cent.	Reduction of Area. Per cent.	Loss of Strength by Welding. Per cent.
13	0.22	2.05	Normal,	85,920	33	59	
			Welded,	80,360	17	25	5.3
14	0.16	3.35	Normal,	97,820	29	56	
			Welded,	87,500	6.9	7.5	10.5
19	0.19	2.62	Normal,	148,100	12.5	17.4	
			Welded,	117,800	1.7	1.7	20.4
24	0.54	3.00	Normal,	151,430	8.0	8.4	
			Welded,	117,600	2.3	1.7	22.3
29	0.96	3.10	Normal,	135,550	8.5	28.9	
			Welded,	114,600	4.6	4.5	15.4
41	0.51	4.93	Normal,	115,730	3.5	6.9	
			Welded,	104,050	3.4	2.9	10.0

carbon, 3 per cent. nickel-steels, the average of which is as follows:

	Ultimate Strength. Lbs. per Sq. In.	Elongation. Per cent.
Normal,	78,622	24
Welded,	78,904	16

These tests show the same strength, but a decrease in ductility after welding.

It may be taken as a rule, however, that steel containing over 3 per cent. of nickel is difficult to weld, the cause being the tenacity with which a film of oxide clings to nickel-steel. A good flux lessens, but does not entirely remove, this difficulty.

The Effect of Nickel on Hardness.

"You cannot make a very low-carbon steel hard by the addition of nickel alone." This opinion, expressed by Mr. Chase, of the Midvale Steel Co., is corroborated by others. Gun-barrels, containing 4.5 per cent. of nickel and 0.30 carbon, are soft and very ductile; the steel can be cut with a knife,* and yet tensile tests show an elastic limit of 80,000 pounds, with 105,000 pounds ultimate strength. The elongation is 25 per cent., and the reduction of area 45 to 50 per cent. This high tensile strength of 80,000 pounds could be attained in simple steel by the use of about 0.70 per cent. of carbon, but at the expense of a hardness and brittleness not shown by the nickel-

---

* Mr. Porter, Bethlehem Steel Co. Private communication.

steel. Nickel-steel rolls, containing 1 per cent. of carbon and 5 of nickel, made by the Reliance Steel Casting Co., "were certainly turned easier than 1 per cent. carbon simple steel."*

Mr. Hadfield says, in the paper already cited:

"While the samples (containing 11.39 and 15.48 per cent. of nickel) present extraordinarily high resistance to compression, they cannot be described as possessing hardness like that of carbon-steel in its hardened condition. . . . . It is doubtful whether any specimen of carbon-steel in its ordinary condition (that is, unhardened) has ever been known to give such a high tenacity, for example, as Specimen I. (0.18 carbon, 11.39 nickel), which, in the unannealed condition, possesses a tensile strength of 94 tons (210,560 pounds) per square inch. Yet these specimens, to the machine-tool or file, in no way compare with the hardness of carbon-steel."

The effect of nickel on hardening is not due to the influence of nickel alone, but to its effect on the carbon in rendering that element more sensitive to heat-treatment. If a steel contains less than 6 per cent. of nickel, the influence of the amount of carbon present on the hardness obtained by water-quenching is very strongly marked. Above 8 per cent. of nickel, the effect of the carbon present seems to be partially masked by the nickel; 18 per cent. nickel-steel is as hard and elastic with 0.30 as with 0.75 carbon. If steel containing 18 per cent. nickel and 0.60 carbon be heated and plunged into water, it will be perceptibly softened; and if the nickel be raised to 25 per cent., this softening is very noticeable.†

"At 20 per cent. of nickel," says Mr. Riley, "successive increments of nickel tend to make the steel softer and more ductile, and even to neutralize the influence of carbon."

An 18 per cent. nickel-steel, with 0.60 carbon, if heated to a good red heat and worked on the anvil until cold, will have the hardness and elasticity, without the brittleness, of a piece of tempered steel. The nickel-steel takes the temper and hardness by hammering alone—water-quenching will soften it. The 18 per cent. nickel-steel is exceedingly hard to punch cold; but, as the nickel increases, the steel softens, and a 25 per cent. nickel-steel can be worked cold almost as easily as German silver.‡ The explanation of this peculiar behavior may be

---

* Mr. Bailey, **Reliance Steel** Casting Co. Private communication.
† Mr. H. J. Williams, Damascus Steel Co. Private communication.
‡ Mr. H. J. Williams, Damascus Steel Co. Private communication.

found in the experiments made in Berlin by Prof. Rudeloff. On melting in crucibles a steel calculated to contain 1 per cent. of carbon and 30 of nickel, he found that 80 per cent. of the carbon present was found in the finished alloy as graphite. With a steel containing 1 per cent. of carbon and 60 of nickel, the same conditions were found to obtain.* This seems to indicate that in steels containing less than 6 per cent. of nickel the effect of nickel was to throw the carbon into a condition where it is exceedingly sensitive to heat-treatment; but as the percentage of nickel rises, the carbon is thrown more and more into the graphitic state, upon which heat-treatment does not produce any effect.

Rigidity of Nickel-Steels.

Under impact-tests, in which a heavy weight is allowed to fall upon a steel bar, supported at its extremities, nickel-steel with about 3 per cent. of nickel shows about 48 per cent. greater stiffness, and 45 per cent. greater toughness, than carbon-steel. The word *stiffness* refers here to the amount of deflection produced by the blow; while the word *toughness* refers to the number of blows required to produce rupture.

Beardmore says that, in experiments with nickel- and carbon-steel of the same carbon-contents, the nickel-steel was fractured at 7 impact-blows, and broke after 35 blows; while the simple steel was fractured at 5 and broke after only 12 blows.

Mr. Taylor, of the Carnegie Steel Co., has kindly furnished copies of impact-tests, made upon simple steel and nickel-steel axles. The weight dropped was 1640 pounds; but as the nickel-steel axles were $1\frac{1}{4}$ inches greater in diameter than those of simple steel, the momentum was made proportional by allowing a drop of 29 feet on the simple steel and of 44 feet on the nickel-steel. In Tables XVI. and XVII., representing these tests, it will be noted that the simple steel was deflected by the first blow 7.5 inches, and broke after 47 blows; while the nickel-steel was deflected only 5.0625 inches, and broke only after 68 blows.

---

* *Verein zur Beförderung des Gewerbfleisses*, 1897, p. 242.

TABLE XVI.—*Record of Tests of Simple Steel Axles, 4⅜ Inches in Diameter at the Center, by the Bureau of Inspection of the Carnegie Steel Co., Ltd.*

Heat. No.	CHEMICAL ANALYSIS.				PHYSICAL TESTS.				Height of Drop. Feet.	Weight of Drop. Lbs.	Deflection after First Blow. In.	No. of Blow at which Axle Broke.	Reduced Diameter of Axle. In.	Remarks.
	C. Per cent	P. Per cent	Mn. Per cent	S. Per cent	Elastic Limit. Lbs. per Sq. In.	Tensile Strength. Lbs. per Sq. In.	Elongation in 2 in. Per cent.	Reduction of Area. Per cent.						
12,170	0.35	0.017	0.50	0.022	49,140	75,190	28.0	44.3	28	1,640	7⅛	48th	4 1/16	Fracture granular. Drop-tests were made on the spring-base machine. These axles were forged from round blooms and re-annealed. No physical tests were made.
11,273	0.35	0.019	0.41	0.022	47,270	71,660	25.0	42.2			6⅛	52d	4 3/16	
12,164	0.40	0.021	0.50	0.031	48,660	74,540	30.0	50.2				41st	4¼	
12,181	0.30	0.014	0.62	0.020	58,760	75,720	26.5	49.4				44th	4 1/16	
12,180	0.37	0.010	0.43	0.021	54,550	72,980	25.0	50.2				59th	4	
11,189	0.36	0.016	0.66	0.022	48,290	76,830	27.0	47.7				47th	4	
12,179	0.35	0.013	0.45	0.022	60,680	71,620	27.0	44.3						
12,269	0.37	0.006	0.45	0.022	44,050	70,610	27.0	42.8						
15,046	0.24	0.009	0.42	0.033					29					
15,046	0.24	0.009	0.42	0.033										
20,040	0.26	0.016	0.44	0.037										
14,651	0.25	0.020	0.44	0.031										
14,651	0.25	0.020	0.44	0.031										

TABLE XVII.—*Record of Tests of Nickel-Steel Axles, 5⅜ Inches in Diameter at the Center, and Tested to Destruction, by the Bureau of Inspection of the Carnegie Steel Co., Ltd.*

(Drop-Tests made on Spring-Base Machine. Height of Drop, 44 Feet; Weight, 1640 Pounds.)

The Effect of Nickel on Compressive Strength.

In an alloy of pure iron with varying percentages of nickel, the resistances to compression have been determined. Resistance to direct compression varies directly with the percentage of nickel in the alloy, up to 16 per cent., very nearly in the same manner as the resistance to tension, except that for tension the maximum strength is reached at 8, while the maximum for compression is reached at 16 per cent. In comparative tests, in which small blocks of the alloys were subjected to the blows of a heavy hammer, the same ratio between the nickel-contents and the resistance to compression was observed; and it was noted also that, up to 16 per cent. of nickel, the resistance of the block to compression increases with the number of blows given, and that this increased resistance is also in direct ratio to the amount of nickel, up to the maximum of 16 per cent. In other words, nickel seems to increase the gain in strength produced by cold-working.*

Mr. Hadfield has published the results of compression-tests on nickel-steels of low carbon.† He finds that samples containing 11.39 and 15.48 per cent. of nickel possess extraordinarily high compressive-strength. These samples showed a shortening of only 1 per cent. by a load of 100 tons; while, under the same conditions, a low-carbon steel, without nickel, was shortened 50 per cent.

TABLE XVIII.—*Compression-Tests by Mr. Hadfield.*

Mark of Test.	Carbon. Per cent.	Nickel. Per cent.	Elastic Limit by Compression. Tons per Sq. In.	Shortening by 100-Ton Load. Per cent.
a	.19	.27	22	50
b	.14	.51	22	50
c	.13	.95	20	49
d	.14	1.92	27	47
e	.19	3.82	28	41
f	.18	5.81	40	37
g	.17	7.65	40	33
h	.16	9.51	70	3
i	.18	11.39	100	1
j	.23	15.48	80	1
k	.19	19.64	80	3
l	.16	24.51	50	16
m	.14	29.07	24	41

* Rüdeloff, *Verein zur Beförderung des Gewerbfleisses*, 1896.
† "Alloys of Iron and Nickel," *Inst. Civil Engrs.*, vol. cxxxviii.

Mr. Hadfield says:

"Some noteworthy results are obtained, especially in samples i and j, in which the resistance to compression is extraordinarily high. These results are the more remarkable in view of the low carbon present. A still more curious point is, that notwithstanding the high resistance to compression stresses, they are comparatively soft and can be machined—though not readily. There is no doubt that as regards pure alloys of iron and nickel, that is, no carbon or other elements being present or to only a small extent, the increase in resistance to stress is a most remarkable property, apparently only possessed by these nickel-iron alloys."

The Shearing-Strength of Nickel-Steels.

In shearing the edges of rolled deck-plates, carbon-steel often cuts raggedly—small fragments breaking back from the edge of the shears. Nickel-steel, on the other hand, cuts clean, like a very mild steel.* This is also true of punching. "Careful experiments in punching show that the hole is left with clean-cut edges, with no slivers or wire-edges. Nickel-plates shear more neatly than carbon-plates."†

Prof. Rudeloff has investigated the influence of nickel upon the shearing-strength of iron, with the following results:‡

TABLE XIX.—*Effect of Nickel on Shearing-Strength.*

Nickel. Per cent.	Shearing-Strength. Lbs. per Sq. In.
0,	39,303
1.0,	40,441
2.0,	47,363
3.0,	47,363
4.0,	48,311
5.0,	52,009
8.0,	61,826
15.5,	95,581
30.0,	50,349
60.0,	53,905
93.0,	50,705
98.0,	49,306

The shearing-strength rapidly rises with increase in the percentage of nickel, up to 16 per cent., at which point the nickel-alloy has 2.5 times the shearing-strength of pure iron.

---

* E. F. Wood, Carnegie Steel Co. Private communication.
† "The Advantages of Nickel-Steel," Commander J. G. Eaton.
‡ M. Rudeloff, "Vierter Bericht des Sonderausschusses für Eisenlegirungen," *Verein zur Bef. des Gewerbfl.*, Berlin, 1896.

Further results on the shearing of nickel-steels are given below, under the head of nickel-steel rivets.

Loss of Strength on Punching.

Tests have been made at the Parkhead Forge rolling-mill, in Glasgow, Scotland, to determine the original strength of nickel-steel boiler-plates, and also the residual strength, after a portion had been removed by punching. Two samples of plate were tested: No. 1 being mild steel, of usual "boiler-plate" quality, and No. 2 a steel somewhat higher in carbon. Each contained the usual 3 per cent. of nickel. The tensile strength of the whole sheets, before punching, averaged as follows:

	Lbs. per Sq. In.
No. 1,	84,672
No. 2,	116,480

Test-pieces of these sheets, about 3.5 inches in width, were taken, and, in the center of each, 1-inch holes were punched. The samples were then subjected to tensile tests, with the following average results:

	Lbs. per Sq. In.
No. 1,	71,523
No. 2,	89,510

The loss of strength by punching was, for No. 1, 15.5 per cent., and for No. 2, 20 per cent. of the original strength. On the thicker sections of No. 1 the loss in strength was only about 10 per cent.

Mr. Beardmore states the loss of strength due to punching in ordinary mild steel to be 33 per cent. of the original strength.

This is, however, higher than is usually reported. Prof. Howe says that it ranges from 9 to 34 per cent.—the average of tests on simple-steel plates showing a little over 20 per cent. loss on punching.

It is evident from this comparison that nickel-steel suffers from punching at least as little as mild boiler-steel. Mr. Beardmore claims that in this regard nickel-steel is much superior to the simple steels.*

---

* "Nickel Steel," *Inst. Engrs. and Shipbuilders of Scotland*, March, 1896; *Industries and Iron*, May, 1896.

The Influence of Nickel on Appearance and Tests.

In handling the charge in the open-hearth furnace, the melter relies for guidance upon the appearance of the fracture of small test-bars, taken from time to time. With simple carbon-steels, a very accurate estimate of the amount of carbon present can be thus obtained. Nickel changes entirely the characteristic appearance of the fracture. Test-bars from a heat of nickel-steel show very pronounced crystallization, like the chill of cast-iron, the crystals meeting sharply at the angles of the test-mould. Nickel-steel has a darker fracture than simple steel; and the furnace-man, judging by the appearance of the crystallization, would be apt to think the carbon was higher than is really the case. If, however, a series of test-bars be taken during a heat of nickel-steel, it will be found that the appearance changes regularly as the carbon drops; and the appearance of the fracture, once learned, forms as safe a criterion of the carbon-contents as is furnished by the corresponding changes in the case of simple steel.

In color-carbon tests on nickel-steels, the chemist is apt to estimate the carbon too low—not because of any color given by the nickel present (which, of itself, does not interfere with the color-test), but by reason of the effect of nickel on the carbon, changing part of it to the hardening-condition, in which form its color-effect is not so visible as that of cement-carbon.*
For accurate comparison of tests, therefore, combustion-determinations of carbon are necessary.

The color of the finished steel becomes lighter with increase in the nickel-contents. The ordinary 3.5 per cent. nickel-steels do not perceptibly differ in appearance from simple steels; at 10 per cent. of nickel, the color is noticeably lighter; and, at 18 per cent., the steel has a soft silvery whiteness; with higher percentages the color seems again to darken, and the 25 and 30 per cent. nickel-steels are duller and less lustrous than the 18 per cent. In texture, moreover, the 18 per cent. nickel-steels appear more smooth and close-grained than iron-alloys containing a larger percentage of nickel.

---

* Mr. Chase, Midvale Steel Works. Private communication.

Segregation in Nickel-Steel.

The various elements which enter into the composition of steels—silicon, sulphur, phosphorus, manganese and carbon—are all thoroughly mixed in the molten steel, and, in the furnace, are evenly distributed throughout the charge. When the steel is cast in the ingot-mould, each of these elements forming its own alloy with iron, and each solidifying at a different temperature and possessing a different specific gravity from the others, begins to move through the fluid steel. As the mass cools, the tendency of these compounds is towards the central and upper part of the ingot. Analysis shows that the center of an ingot contains a much greater proportion of carbon, silicon, sulphur and phosphorus than the outside; and for this reason, in the manufacture of shafting, gun-hoops and other articles in which uniformity of composition is a *sine qua non*, the center of the ingot is removed by boring.

"Nickel is supposed to lessen the segregation and liquation of carbon by combining with the carbon which cements the particles of iron together, and thus bringing the specific gravity of the carbon compounds nearer to that of the rest of the alloys in the fluid mixture. It also seems to cause this cementing carbon to solidify at more nearly the same temperature as the other alloys."*

Analyses from an ingot of uncompressed nickel-steel, 66.5 inches in diameter and 122 inches long, made by the Bethlehem Steel Co., gave the following result:†

TABLE XX.—*Segregation in Nickel-Steel.*

	Outside. Per cent.	Six In. from Center. Per cent.	Ratio of Concentration.
Nickel,	3.07	3.27	100 : 108
Silicon,	0.172	0.170	100 : 98
Sulphur,	0.03	0.06	100 : 200
Carbon,	0.31	0.36	100 : 116
Phosphorus,	0.025	0.047	100 : 188

It will be noted that in this nickel-steel ingot the segregation-excess of carbon is only 0.008 per cent. Prof. Howe in his *Metallurgy of Steel*, states that in simple steels the average segregation-excess of carbon is greater than that of any other element; and, in five cases which he cites, it exceeds 0.44 per cent.

---

* Mr. G. H. Chase, Midvale Steel Works. Private communication.
† Mr. H. F. Porter, Bethlehem Steel Works. Private communication.

Coefficient of Expansion.

Simple steel, when heated or cooled 1 degree Centigrade, at ordinary temperature, expands or contracts about 11.6 parts in a million. Expressed in the usual way, the coefficient of expansion of simple steel for each degree C. is 0.0000116.

According to M. Ch. Ed. Guillaume,[*] the coefficients of nickel-steels are normal up to 19 per cent. of nickel, from which point the coefficients rise rapidly up to 24 per cent., then fall to a minimum with steels containing 36 per cent. of nickel, and then, with further increase in nickel, return to a normal value. In investigating the behavior of these alloys, M. Guillaume discovered several remarkable anomalies which led him to divide

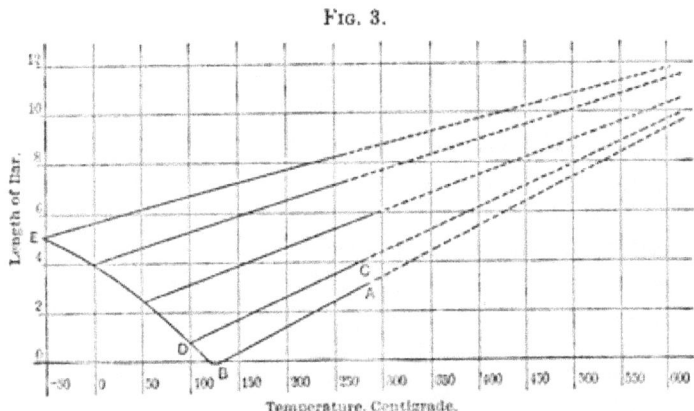

FIG. 3.

EXPANSION AND CONTRACTION OF 15 PER CENT NICKEL-STEEL
BY CH. ED. GUILLAUME.

nickel-steels into two classes: those containing less than 26 per cent. of nickel, which he calls *irreversible* alloys, and which show entirely different behavior at ascending and at descending temperatures; and those containing over 26 per cent. of nickel (the *reversible* alloys), which have a regular coefficient of expansion, the same for rising as for falling temperatures.

All the *irreversible* alloys show practically the same behavior as the 15 per cent. nickel-steel, as shown in Fig. 3, in which the abscissæ represent the temperature and the ordinates the length of the bar. When, after heating to cherry-red, the bar is allowed to cool, the metal contracts regularly, as shown by

---

[*] "Récherches sur les Aciers au Nickel," *Comptes Rendus*, March 11, 1898.

the line A B. At about 130° C., however, the bar commences to expand, and continues to expand as the temperature falls; and the rate of expansion on cooling is 40 parts in a million per degree C. If now, while the bar is cooling and expanding, it be taken (say at 100° C., when it has expanded to the point D) and reheated, it does not (as would be expected) contract till it reaches a temperature of 130°, and then expand, but expands immediately, along the line C D. On cooling, it contracts again along the line C D, and then expands along the line D E, continuing to expand as the temperature falls, until, at 60° below zero, C., the expansion is very slight. There are, for these "irreversible" alloys, two transformation-points; one near a red heat and the other at about 60° below zero, C. The more nickel contained in the steel (up to 26 per cent.), the lower the point (B), at which transformation by cooling commences; and the more rapidly is the point E attained, at which transformation is complete. The coefficient of expansion on heating from — 60° is 0.00001047, or 10.47 parts per million for each degree Centigrade.

The *reversible* nickel-steels (those containing over 26 per cent. of nickel) possess a regular coefficient of expansion, which varies according to the percentage of nickel. These coefficients, as given by M. Guillaume,* are as follows:

TABLE XXI.—*Coefficient of Expansion of Nickel-Steels.*

Nickel. Per cent.	Coefficents of Expansion between 0 and θ (in Centigrade Degrees).
26.2,	$(13.103 + 0.02123\,\theta)\ 10^{-6}$
27.9,	$(11.283 + 0.02889\,\theta)$ "
28.7,	$(10.387 + 0.03004\,\theta)$ "
30.4,	$(\ 4.570 + 0.01194\,\theta)$ "
31.4,	$(\ 3.395 + 0.00885\,\theta)$ "
34.6,	$(\ 1.373 + 0.00237\,\theta)$ "
35.6,	$(\ 0.877 + 0.00127\,\theta)$ "
37.3,	$(\ 3.457 - 0.00647\,\theta)$ "
39.4,	$(\ 5.357 - 0.00448\,\theta)$ "
44.4,	$(\ 8.508 - 0.00251\,\theta)$ "
100.0,	$(12.514 + 0.00674\,\theta)$ "

The 26 per cent. nickel-steels expand a trifle more than pure nickel; then the coefficient of expansion rapidly decreases, reaching its minimum at 26 per cent. nickel, where the alloy

---

* *Comptes Rendus*, March 11, 1898.

expands by heating only one-tenth as much as platinum and one-twentieth as much as brass.

Calculating from the above formula, it will be seen that for each degree Centigrade, nickel-steel of 36 per cent. nickel expands 0.878 parts in a million. Platinum expands 8.84, and brass 18.78 parts, under the same conditions.*

Density of Nickel-Steels.

The specific gravity of various nickel-iron alloys has been reported by various investigators as follows:

TABLE XXII.—*Specific Gravity of Nickel-Iron Alloys.*

Carbon.	Nickel. Per cent.	Character.	Density.	Observer.
Simple steel......	0.		7.6 to 8.0	
Low-carbon......	3.0	Forged.	7.862	Beardmore.
0.19-carbon......	3.82	Cast.	7.777	Hadfield.
	5.0	Forged.	7.787	Guillaume.
0.17-carbon......	7.65	Cast.	7.743	Hadfield.
Low-carbon......	8.0	"	7.892	Rudeloff.
	10.0		7.866	Riley.
	15.	Forged.	7.903	Guillaume.
0.23-carbon......	15.48	Cast.	7.752	Hadfield.
	19.0	Forged.	7.913	Guillaume.
	24.0	(Magnetic.)	8.014	"
	24.	(Non-magnetic.)	8.111	"
0.16-carbon......	24.51	Cast.	7.810	Hadfield.
	26.		8.08	Riley.
	26.2	Forged.	8.096	Guillaume.
	29.	Wire.	8.4	Von Helmholz.
0.14-carbon......	29.07	Cast.	8.054	Hadfield.
	30.4	Forged.	8.049	Guillaume.
	31.4	"	8.008	"
	34.6		8.066	"
	37.2		8.005	"
	39.4		8.076	"
	44.3		8.120	"
	98.8	(Commercial Cast.)	8.839	Hadfield.
	98.8	" Forged.)	8.826	"
	98.8	" "	8.86	Riley.

Corrodibility of Nickel-Steel.

Nickel itself is very slightly affected by air, steam, moisture and acid vapors; in alloys with iron and steel, the incorrodibility increases with the percentage of nickel up to 18. "This alloy is practically non-corrodible. A sample hung for three months in the laboratory-fumes was very slightly corroded in spots, which seemed due to the condensation of acid in drops

---

* Thurston's *Materials of Engineering.*

on the surface."* Further increase in the percentage of nickel causes steel to become open-grained and not susceptible to tempering, and (probably for this reason) the 25 per cent. nickel-alloy will gather a faint coating of rust, where the 18 per cent. alloy would remain bright. "Tico" resistance-wire, containing 27.5 per cent. of nickel, was very slightly coated with rust after one year's exposure to water in a very wet cellar; iron wire, under the same conditions, was entirely changed to oxide. With the ordinary nickel-steels (3 to 3.5 per cent. nickel) corrosion is slightly less than in simple steels. Mr. Whyte, superintendent of Lieth docks, has tested the loss in weight of nickel-steel, mild steel and wrought-iron, after twelve months' exposure to sea-water, with the following results:†

TABLE XXIII.—*Corrosion of Nickel-Steel, etc., by Sea-Water.*

	Loss in Weight. Per cent.
Planed nickel-steel (3 per cent. Ni),	1.36
" simple " (0.25 per cent. C),	1.72
" wrought-iron,	1.89
Unplaned nickel-steel,	0.74

Riley states that, as compared with 0.18 carbon-steel, the loss in weight by immersion in dilute hydrochloric acid was as follows:

TABLE XXIV.—*Corrosion of Nickel-Steel in Dilute Hydrochloric Acid.*

	Parts Actual Loss in Weight.
0.18-carbon steel,	1,000
5 per cent. nickel-steel,	830
25 " "	1.15

Wiggin gives the percentage of loss in steam and salt water as follows:

TABLE XXV.—*Corrosion of Nickel-Steel, etc., by Steam and Salt Water.*

	Loss by Boiling for 3 months in water with 10 Per cent Salt. Per cent.	Loss by Treating with Steam for 2 months. Per cent.
Nickel-steel (3 per cent.),	1.00	.27
Bessemer steel,	1.81	.58
Open-hearth steel,	1.97	.31
" "	2.00	.36

* H. J. Williams, Damascus Tool-Steel Co., Pittsburgh.
† Beardmore, "Nickel-Steel as an Improved Material for Boiler-Plates."

Prof. Howe states that on exposure to fresh water and to air he has found the average loss in grammes as follows:

TABLE XXVI.—*Loss of Weight of Nickel-Steel, etc., in Fresh Water and Air.*

	Immersed in Fresh Water. Grammes.	Exposed to Air. Grammes.
High-nickel steel,	79	29
Low- "	251	61
Wrought-iron,	279	87
Soft steel,	274	117

Mr. Hadfield, experimenting on forged test-bars of nickel-steel containing about 0.20 per cent. of carbon, immersed for three weeks in 50 per cent. sulphuric acid, finds less difference in corrosion between nickel-steel and simple steel than is found by other investigators.

TABLE XXVII.—*Loss of Weight through Prolonged Immersion in Sulphuric Acid.*

Analysis.		Loss.
Carbon. Per cent.	Nickel. Per cent.	Per cent.
0.13	0.95	3.23
0.19	3.82	2.95
0.17	7.62	2.77
0.18	11.39	2.60
0.14	29.07	1.34

Experiments made by the writer seem to prove that the amount of carbon present has much influence on the corrodibility of nickel-steel. These tests were made by placing bars of the steel to be tested in a solution of 10 per cent. hydrochloric acid and 10 per cent. common salt in water. The samples were held apart by strips of wood, to avoid electrolytic action. The pickling was continued for a week; the samples were then carefully washed, brushed and weighed. The results are given in full in Table XXVIII. Arranging the samples into the three groups,—low, medium and high carbon,—it will be noted that in each group there is a fairly regular decrease in corrosion, corresponding to increase in nickel. Among special steels, that containing 18 per cent. of nickel gives remarkable results, showing a loss by corrosion of only 0.077 per cent., against a loss of 5.8 per cent. in low-carbon steel containing no nickel.

TABLE XXVIII.—*Corrosion-Tests of Nickel-Steels by Treatment with Hydrochloric Acid and Brine.*

Carbon. Per cent.	Nickel. Per cent.	Loss. Per cent.
0.19	0.00	5.811
0.19	1.05	1.305
0.19	2.62	1.712
0.15	3.17	.664
0.16	3.55	.429
0.18	6.30	.547
0.33	0.00	2.121
0.37	1.05	1.287
0.29	1.90	.607
0.31	3.40	.434
0.31	3.75	.345
0.43	5.75	.318
0.73	0.00	1.508
0.81	1.15	.914
0.81	2.30	.924
0.91	3.10	.540
1.00	3.52	.483
.....	11.25	1.56
0.45	18.	.077
.....	23.86	.479
.....	99.8	.130

From the above tests it appears that, with steel of the same carbon-contents, corrosion decreases as nickel increases, while carbon itself, irrespective of the nickel present, seems also to increase the resistance to corrosion.

Electrical and Magnetic Qualities.

TABLE XXIX.—*Specific Resistance of Nickel-Steels.*

	Specific Resistance in Ohms per Meter 1mm. sq.	Authority.
Pure nickel,	0.126	Matthiessen.
Pure iron,	0.098	Matthiessen.
Nickel-steel,	0.80 to 0.90	Guillaume.
"Tico" wire,	0.810	Trenton Iron Co.
"Superior" wire, 86 microhms at 20° C.	0.860	Von Helmholz.
"Climax" wire,	0.784	Roebling.
German silver,	0.338	Von Helmholz.

According to Guillaume, all nickel-steels have a high electrical resistance which does not seem to vary very much with the percentage of nickel. The nickel-steel resistance-wires, "Tico," "Superior" and "Climax," containing from 25 to 30 per cent. of nickel, have about 48 times, while German silver has about 18 times, the resistance of copper.

Mr. Hadfield reports a very interesting series of tests of the electrical conductivity of nickel-steels, as follows:

TABLE XXX.—*Electrical Conductivity of Nickel-Steels.*

ANALYSIS.		Condition of Material.	COMPARATIVE ELECTRIC CONDUCTIVITY.	
Carbon. Per cent.	Nickel. Per cent.		Copper 100.	Iron 100.
0.14	1.92	Unannealed.	8.00	51.5
0.19	3.82	"	6.88	44.3
0.18	11.39	"	4.49	28.9
0.19	19.64	"	4.00	25.7
0.16	24.51	"	2.67	17.2
0.16	24.51	"	3.70	23.84
0.16	24.51	Water quenched.	3.02	19.45
......	31.00	Unannealed.	......	12.02

Hopkinson investigated, in 1890, the magnetic properties of nickel-steels. In examining a steel containing 4.7 per cent. of nickel and 0.22 of carbon, he found that on heating to 750° C. the steel became non-magnetic and absorbed considerable heat. On cooling to 632° C., this latent heat was given out and the metal again became magnetic. Nickel-steel with 24.5 per cent. of nickel was non-magnetic at ordinary temperatures, and remained non-magnetic up to 700° C. This material did not recalesce on cooling. If cooled down a little below the freezing-point, this steel became magnetic and remained so until heated up to 580° C. At this temperature it became non-magnetic, and remained so on cooling to the temperature of the air. From 4° C. to 580° C., this alloy can exist in two states: either magnetic or non-magnetic, according to the previous heat-treatment. The specific electrical resistance is .000052 in the magnetic and .000072 in the non-magnetic state.*

According to Guillaume, all nickel-steels below 25.7 per cent. Ni can be, at the same temperature, either magnetic or non-magnetic, according to their previous heat-treatment. These steels are comprised between the formulæ Fe and Fe_3Ni, and they show different properties at ascending and at descending temperatures.†

---

* Hopkinson, "Magnetic Properties of Nickel-Iron Alloys," *Proc. Royal Soc.*, 1890; *Engineering*, Sept. 5, 1890.
† "Récherches sur les aciers au Nickel," *Comptes Rendus*, March, 1898.

When these nickel-steels, containing less than 25 per cent. Ni, are heated, they lose their magnetism at a point between dull and cherry-red. When they are cooled, they do not become magnetic until a temperature is reached much lower than that at which magnetism was lost. The return to the magnetic state is gradual; and the point at which magnetism is recovered is lower, in proportion, as the alloy is richer in nickel. For an alloy of 24 per cent., the return to the magnetic state occurs a little below zero. The alloys increase in volume on becoming magnetic, the density of the 24 per cent. nickel-steel being 8.111 in the non-magnetic, and 8.014 in the magnetic state. The temperature at which the loss of magnetism occurs can be represented as a function of the nickel present by the formula:

$$\theta = 34.1\ (n-26.7) - 0.80\ (n-26.7)^2,$$

θ representing the temperature and n the percentage of nickel.

Steels containing more than 25 per cent. of nickel possess regular magnetic properties which depend on the actual, and not on the previous, temperature.

M. Marcel Deprez,[*] commenting on M. Guillaume's equation representing the loss of magnetism at given temperatures corresponding to the amount of nickel present, shows that for alloys with 26.7 per cent. the transformation takes place at 0° C.; while for 39.4 per cent. Ni, the temperature is 315°, and for 58 per cent. the highest value, 363° C., is attained. The transition from magnetic to non-magnetic condition takes place within a range of 50°. To eliminate the magnetism by means of boiling water, a 30 per cent. alloy is needed. This metal is strongly magnetic at 50° C.

The lower nickel-steels, containing 3 to 5 per cent. of nickel, possess a magnetic permeability greater than that of wrought-iron.

The magnetic saturation of simple steels and nickel-steels, at ordinary and at exceedingly low temperatures, has been investigated by Prof. Dewar. The method followed was to charge small magnets to saturation at the ordinary temperature, measure the deflection produced on a galvanometer, then plunge the

---

[*] *Rev. Générale des Sciences*, Feb. 15, 1898; *Jour. Iron and Steel Inst.*, 1898, p. 506.

magnets into liquid air, whereby a temperature of —186° C. is attained, and note the magnetic strength again.

Simple steels behaved in the same general way; the first cooling diminished the magnetic moment 6 per cent., and on heating up to ordinary temperature, it was still further diminished about 16 per cent. On cooling again the magnetic moment increased 6 per cent., and from this point heating always diminished and cooling always increased the magnetic moment; so that at —186° C. the magnetism was 10 per cent. greater than at 5° C. The increase of magnetic moment, when cold, varies with the amount of carbon; hard steels showing 10, medium steels 20, and soft steels 30 per cent. increase on cooling.

With nickel-steels a different result was obtained. The steels containing 0.94, 3.82 and 7.65 per cent. of nickel all acted like high-carbon steels, the magnetic moment being 10 to 11 per cent. greater when cold than at ordinary temperatures. The alloy with 19.64 per cent. nickel showed a very considerable loss—nearly 50 per cent.—of magnetic moment on first cooling; on removal from the liquid air, the magnetism increased; and after that the magnetic moment was 25 per cent. smaller when cold than at ordinary temperatures.*

III.—MOLECULAR RELATION OF NICKEL TO IRON AND CARBON IN NICKEL-STEELS.

It seems to be a generally accepted opinion that the effect of nickel in nickel-steels is two-fold:

1. With low-carbon, or nearly pure wrought-iron, nickel seems to form a homogeneous alloy, much tougher and stronger than either pure nickel or pure iron. This conclusion is strengthened by the experiments of M. Moulan, referred to on page 15, and by the experiments of Rudeloff, on page 6, of this paper.

2. It is well known that pure nickel and iron unite to form definite crystalline compounds. Such an alloy, containing iron 45.64, nickel 54.36, and corresponding to the formula Fe_8Ni_9, was discovered by the writer in nickel-matte and described in

---

* Dewar and Fleming on "Changes Produced in Magnetized Iron and Steel by Cooling," *Trans. Royal Soc.*, May 21, 1896.

the *Journal of Analytical and Applied Chemistry*, March, 1892. It occurs in the shape of thin, flat crystals, looking like corner-clippings of tin foil, and having the form of right-angled isosceles triangles, measuring at the most one-half inch on the longer side. The crystals are silver-white, exceedingly tough, flexible and malleable, and strongly magnetic. As far as can be judged they are as soft as pure iron, if not softer.

M. Guillaume, in his researches on the magnetic properties and expansion-coefficients of nickel-steels, finds that the most pronounced results and most remarkable peculiarities are shown by the alloys which correspond most closely to the formulæ Fe_2Ni and Fe_3Ni.

It would seem from this that nickel forms with iron a series of crystalline chemical compounds, which, from the experiments of Moulan and Rudeloff, would seem to be tougher and stronger than pure iron.

Assuming that, as many hold, nickel causes carbon to pass from the cement- to the hardening-condition, we may conceive that the stiffness and rigidity of nickel-steels are due to the effect of carbon in this condition; while the ductility of nickel-steel is due to the fact that the matrix (the nickel-iron "mineral") is tougher than simple iron, and permits more molecular distortion before breaking.

If we consider the effect of nickel in Harveyized armor-plate we see that in the unhardened under-part of the plate the nickel-iron alloy is soft, tough and ductile; while in the hardened surface the high-carbon nickel-steel is stiff, rigid and hard. Owing to the toughness of the back, a blow on the face is absorbed and its momentum used up in heating the whole plate, while in the case of simple steel a blow expends its momentum in local shattering of the surface. We consider, therefore, that the benefits arising from the use of nickel in armor are two-fold: it forms on the surface a hard, impenetrable carbon-alloy, back of which, and supporting it, is a soft, ductile, homogeneous base. We may conceive of the action of nickel in nickel-steel in much the same way, and consider the stiffness and rigidity of the metal as due to the combination of iron with carbon in the hardening condition, and the ductility of nickel-steel as due to the toughness and strength of the original matrix or crystalline compound of nickel and iron.

NICKEL-STEEL; A SYNOPSIS OF EXPERIMENT AND OPINION. 43

As previously stated, small amounts of nickel seem to throw carbon into the hardening-condition, and large amounts seem to cause the separation of carbon as graphite. The effect of nickel on carbon is well brought out by melting nickel with pig-iron. Dr. S. H. Emmens found that gray pig-iron, melted with 10 per cent. of nickel under charcoal, gave an intensely hard alloy, which could not be machined, but which was nevertheless gray, crystalline in fracture, and very strong. Mr. L. Dunham, Supt. of the Ashland furnaces, states that he has mixed nickel with cast-iron and finds that it will turn white iron to gray.

Some experiments made for the Canadian Copper Co. by Mr. Crowell, of Cleveland, confirm these statements. No. 2 charcoal pig-iron was melted, with varying amounts of nickel, and cast in sand-moulds into bars 1 inch square by 1 foot long. These bars were tested by transverse strain, being supported on bearings 7 inches apart. Four bars of pig-iron, free from nickel, broke under an average load of 4530 pounds. Iron, alloyed with 0.75 per cent. of nickel, showed a strength of 4400 pounds. One per cent. of nickel gave 4175 pounds; 2 per cent., 4800 pounds, and 3 per cent. 5200 pounds, as breaking-load. The fractures were all of good gray surface. The gain of strength due to 3 per cent. of nickel was only 12 per cent. To test the effect on chill, two castings were made in sand-moulds having one steel face; one casting of No. 2 pig-iron was chilled white throughout, while the other casting, of the same pig-iron with 1 per cent. of nickel, was chilled only one-third; indicating that nickel has a retarding effect on the chill.

It is somewhat curious that the presence of less than 1 per cent. of nickel, whether in pure iron, steel or cast-iron, seems to produce a detrimental effect. In all of Rudeloff's experiments, the alloys with 0.5 per cent. of nickel, whether tested by tension or compression, were decidedly lower in elastic limit and ultimate strength than pure iron. In Mr. Hadfield's experiments, low-carbon nickel-alloys with 0.5 per cent. of nickel show a perceptible loss of strength. In a series of experiments made by the Canadian Copper Co. it was noticed that alloys with less than 1 per cent. of nickel were decidedly weaker than the same steel without nickel; and the only plausible explanation was that in small amounts, under 1 per cent., nickel seemed to be a foreign element, producing weakness; while in larger amounts, over 2 per cent., it seemed to unite with the steel and form a tough, homogeneous alloy.

IV.—Uses of Nickel-Steel.

The greater portion of the nickel-steel produced at the present time contains from 0.20 to 0.40 carbon and from 3 to 5 per cent. nickel. As this metal retains all the ductility of a low-carbon, and attains at the same time the strength and stiffness of a high-carbon steel, it has found a great variety of uses. In the following pages the opinions of steel-makers and steel-users, in regard to the application of nickel-steel to various purposes, will be given, as nearly as possible, in their own words.

Boilers.

Mr. Wm. Beardmore, speaking before the Institute of Engineers and Shipbuilders of Scotland, in 1896, said:

"With the growing demand for ships to attain a speed of 30 knots, boilers to work at higher pressure, greater strength with lighter section, it would seem as if a better iron were needed to supply the latest wants of engineers and shipbuilders. We require a metal which can be worked without any special care on the part of the artisan; a metal which in ship-building will enable us to reduce the scantlings, take from the weight of the boilers and add to the strength and reliableness of the propeller shafts. Nickel-steel fulfills all these conditions, and is, in my opinion, a most suitable metal with which to meet the demands for a metal stronger than steel. In nickel steel of .26 carbon we have a metal whose elastic limit is equal to the ultimate strength of ordinary carbon-steel. Mild nickel-steel gives all the properties of high carbon metal, without the treacherous brittleness so painfully evident in the latter."*

At the Cleveland Rolling Mill Co. the effect of 2.7 per cent. of nickel in flange-steel of 0.10 carbon was tested; the two metals, practically identical except as to nickel, gave the following average results:

	Elastic Limit. Lbs. per Sq. In.	Ultimate Strength. Lbs. per Sq. In.	Elongation. Per cent.	Reduction of Area. Per cent.
Flange-steel, 0.10 C.,	35,000	54,000	27.4	54
" " 0.08 C., 2.69 Ni.,	47,000	65,700	24.7	52

The addition of 2.69 per cent. of nickel to this grade of steel increased the elastic limit 31 per cent. and the ultimate strength

---

* Beardmore, *Industries and Iron*, May, 1896; *Proc. Inst. of Nav. Architects*, April, 1897.

20 per cent., while the ductility (as shown by elongation) is nearly the same in both steels.*

Experiments in flanging nickel-steel plates of every thickness suitable for boilers show that this material is worked without any difficulty; it can be readily forged and pressed into dies, without cracking, and its large elongation enables it to be worked to great advantage.† Comparative tests of boiler-plate for the United States vessel Chicago, with a boiler-plate made from the same metal without nickel, gave the following results:

	Elastic Limit. Lbs. per Sq. In.	Ultimate Strength. Lbs. per Sq. In.	Elongation. Per cent.	Reduction of Area. Per cent.
Nickel-steel boiler-plate,	61,830	87,380	21.7	52.2
Carbon-steel,	31,840	60,650	27.0	49.5

Mr. Landis says that when the boiler is made entirely of nickel-steel, thus preventing electrolytic action, there is no doubt but that it is the best material yet applied to this purpose.‡

Shafting.

Owing to the great length and high speed of modern steamships, it is necessary to make use of exceedingly high-grade material in crank- and connecting-shafts.

The U. S. government requires that shafts for the navy be hollow; and this custom is being rapidly taken up in general marine practice. A hollow-forged oil-tempered nickel-steel shaft has been made for the United States battleship Brooklyn: outside diameter, 17½ inches; inside, 11 inches; length, 38 feet 11¾ inches; weight, 19,112 pounds. Test-bars cut from this shaft gave a tensile strength of 94,245 pounds per square inch; elastic limit, 60,770 pounds; elongation, 25.55 per cent.; and reduction of area, 60.58 per cent.

Prof. Merriman is quoted in a paper read before the Society of Naval Architects and Marine Engineers, in 1893, by R. G. Davenport, as estimating the strength of these shafts, compared with solid shafts, when strained to one-half of their elastic limit, as follows:

---

* Sperry. "Nickel and Nickel-Steel," *Trans. A. I. M. E.*, xxv., 63 (March, 1895).

† H. Wiggin, *Iron and Steel Inst.*, Aug., 1895.

‡ "Nickel-Steel," *Scientific American*, Jan. 9, 1897.

Propeller shaft, United States battleship Brooklyn, nickel-steel:
  (a) Horse-power transmitted, at 50 revolutions per minute, 15,780.
  (b) Load, in pounds, at middle of a span of 12 feet on two supports, 276,200.

Simple-steel shaft, solid, 13 inches in diameter (same weight as above):
  (a) Horse-power, transmitted under similar conditions, 5130.
  (b) Load, in pounds, under similar conditions, 89,000; comparative strength, 3 to 1.

"A solid shaft of simple steel of the same strength as the nickel-steel shaft would be 18.9 inches diameter and weight 53 per cent. more."*

The shafting for the battleship Iowa is of nickel-steel containing 0.27 carbon and 3.19 nickel. An average of 48 bars cut from this shaft gave the following tests:

Elastic Limit. Lbs. per Sq. In.	Ultimate Strength. Lbs. per Sq. In.	Elongation. Per cent.	Reduction of Area. Per cent.
60,000	94,500	24.6	59.5

The great gain in elastic limit permits reduction in section, with consequent saving of weight in heavy moving parts of high-speed engines.†

The American liner Paris has had constructed for her a spare length of shafting of nickel-steel, which has a tensile strength of 90,000 pounds per square inch, probably 25,000 pounds more than any British or German shaft.‡

The nickel-steel shaft for the North German Lloyd steamer Kaiser Wilhelm der Grosse weighs 83.3 metric tons (183,593 pounds), is 45 feet 9½ inches long, and shows a tensile strength of 39.4 tons (88,256 pounds), with 20 per cent. elongation.§

Mr. Beardmore‖ points out that a very striking feature of nickel-steel is, that a crack appearing in it will not develop, as in carbon-steel. One of the most frequent causes of casualties at sea, he says, is the breaking of propeller-shafts, due to the development of some flaw in the shaft. Having a number of bars made, 1½ inches square by 18 inches long, of nickel-steel and of ordinary steel, of the same carbon-content, he subjected them to the usual fatigue-tests. The carbon-steel was fractured

---

* "Fatigue of Metals," H. F. J. Porter, Franklin Inst., Dec., 1897.
† "The Advantage of Nickel-Steel over Carbon-Steels," Commander J. G. Eaton, U. S. N.   ‡ *The Steamship*, June, 1894.
§ *Stahl und Eisen.*, vol. xvii., pp. 484–485.
‖ Beardmore, *Inst. Nav. Arch.*, April, 1897.

after 5 blows and broke after 12 blows, while it required 7 blows to fracture the nickel-steel and 35 blows to break it. He says that nickel-steel tears gradually, while carbon-steel, once cracked, breaks short, and adds that, in his opinion, if propeller-shafts were made of nickel-steel, the question of failures would seldom or never be raised, for the reason that, should a crack appear at all in nickel-steel, it will not develop as it would in ordinary carbon-steel.

Engine-Forgings.

In discussing recent improvements in the manufacture of engine forgings Mr. Davenport said, in 1895 :*

"In the highest development of the modern marine engine, reduction of weight of all parts is of prime importance. A very pronounced improvement in strength and toughness can be obtained by the use of nickel-steel. The use of nickel allows a reduction of carbon, makes the steel more sensitive to temper, and facilitates the tempering of irregular shapes. In cases where, owing to thickness of sections and irregular shapes, tempering is not advisable, nickel-steel will show a higher combination of elasticity and toughness than any other material known."

This opinion was corroborated in 1897 by the *Railway Gazette*, which said :†

"It is conceded by the best authorities that from nickel-steel, properly worked, forgings of such parts (locomotive cross-head pins, crank-pins, piston-rods and driving-axles) can now be made which greatly excel similar forgings of all other materials in elastic strength and toughness."

Mr. Porter, of the Bethlehem Steel Company, discussing this topic before the Western Society of Engineers, observed :‡

"Breakages of crank-pins and piston-rods of locomotives have caused mechanical men to seek a metal of high elastic limit and elongation which will successfully resist the severe alternating stresses to which they are subjected. . . . . Mild steel for connecting-rods should contain 0.20 to 0.25 carbon, and show, in specimens four diameters in length, a tensile strength of not less than 57,000 pounds per square inch, and an elastic limit of not less than 27,000 pounds, with an average elongation of 25 per cent.

"For the general run of engine-forgings, where little machine work is required, it is advisable to use a higher-carbon steel, with a tensile strength of about 75,000 pounds per square inch and an elastic limit of 35,000 pounds per square inch, together with an average elongation of 20 per cent. in four diameters.

"For such parts as crank-pins, cross-head pins and parts of machinery subject to severe alternating stresses and wearing action, a still higher grade of steel is

---

* R. G. Davenport, *Cassier's Mag.*, August, 1895.
† *Railroad Gazette*, December 17, 1897.
‡ H. F. J. Porter, *Western Soc. of Engrs.*, October 7, 1896; *Railroad Gazette*, October, 1896.

recommended. Such steel should have a tensile strength of about 85,000 pounds, an elastic limit of about 40,000 pounds per square inch, and an elongation of 15 per cent. in four diameters. If such forgings are tempered, the tensile strength will become about 85,000 to 90,000 pounds per square inch; elastic limit, 45,000 to 55,000 pounds per square inch, and elongation 15 to 20 per cent. By introducing about 3 per cent. of nickel into steel, a finely-granular condition results and a high quality of steel is obtained. By hollow-forging nickel-steel, a material is obtained excelling all others known in elastic strength and toughness. Such forgings were used for the crank-pins, cross-head pins and axles of the new Purdue locomotive, Schenectady No. 2, while the piston-rods were of the same material, but forged solid. Test-bars of this material, ½-inch in diameter and 2 inches between measuring points, showed the following physical characteristics:

TABLE XXXI.—*Test of Nickel-Steel Forgings.*

Oil-Tempered, Annealed Nickel-Steel.	0.25 C., 3.5 Ni. Pounds per Square Inch.
Tensile strength,	91,000
Elastic limit,	57,000
	Per cent.
Proportion elastic to tensile,	62.4
Elongation,	25.05
Contraction,	56.45

"If we represent the strength of a solid wrought-iron shaft 14 inches in diameter and 30 feet in length, as shown in the upper figure, by the figure 1, a solid steel shaft of the same dimensions would be represented by the figures 1.29; if we were to make it of nickel-steel, its strength would be represented by 2.6. Now, if we were to take the same iron shaft and simply bore and anneal it, putting a 3½-inch hole through it, its strength would be represented by 1, just the same as the upper shaft; if we subsequently oil-temper it, its strength would be 1.89. A hollow forged-steel shaft of the same weight as the first, but of 22-inch outside diameter, with a 17-inch hole through it, would be represented by the figure 4; if oil-tempered, 5½; if made of nickel-steel, its strength would be represented by the figure 6, and if oil-tempered, by the figure 8."*

Speaking of the use of nickel-steel shafting in torpedo-boats, where reduction of weight is demanded, Mr. Davenport says:†

"The possible reduction in the weight of shafting for torpedo-boats by the use of steel of high elastic limit is exemplified in the case of the shafting made by the Bethlehem Iron Co. for the Herreschoff Manufacturing Co., to be used in Torpedo Boats 6 and 7. The first inquiry called for solid shafts, about 6 inches diameter, of 80,000 pounds tensile strength and 26 per cent. elongation in 2 inches by ½ inch diameter. To meet these requirements it was proposed to use nickel-steel annealed, but not oil-tempered, of which the elastic limit would be about 50,000 pounds per square inch. It was pointed out by the manufacturers that if the shaft was bored and tempered, an elastic limit of 65,000 pounds, with an elongation of 22 per cent., could be guaranteed, and that if the elastic limit were

---

* H. F. J. Porter, "Fatigue of Metals," *Jour. Frank. Inst.*, December 7, 1897.
† "Steel for Marine Forgings," *Cassier's Mag.*, August, 1897.

raised from 50,000 to 65,000 pounds, a hole 4.16 inches in diameter could be bored through a shaft 6 inches in diameter, by which its weight could be reduced one-half without reducing its torsional strength.

"The shafts were made of nickel-steel, 6 **inches in** outside and 4 inches inside diameter, tempered and **closed** in at both ends. The average physical qualities shown by official specimens cut from these shafts were:

Tensile Strength. Pounds.	Elastic Limit. Pounds.	Elongation. Per cent.
101,000	68,700	22.12."

The alternating stresses to which the moving parts of engines are subjected, stresses much short of the elastic limit of the metal, will in time cause a molecular fatigue which gives rise to rupture. On this point Mr. H. K. Landis says:

"Where a steel of 0.20 per cent. carbon will withstand 300,000 double **stresses**, on an alternate stress machine, a 0.50 carbon-steel will probably break at 400,000, and nickel-steel at 1,500,000 to 2,050,000 such double stresses, each **stress being two thirds of** its ultimate or very near its elastic limit."[*]

In complicated engine-forgings it frequently happens that, owing to the shape or size of the article, oil-tempering is impracticable. With reference to such contingencies, Mr. Colby[†] observes that in cases where oil-tempering is not practicable and special requirements are demanded, they can be obtained by using a somewhat softer and tougher steel, and the introduction of from 3 to 4 per cent. of nickel.

In the same connection the observation of Mr. Hadfield is of interest, that the addition of nickel, either by conferring greater homogeneity, or by some particular combination with the iron or carbon present, or both, appears to confer properties upon the alloy equivalent to an annealing; or, if annealing be employed, to reduce the stress produced by forging; it does this without injury or seriously lowering the elastic limit.

These reasons have led to the adoption, by a number of railways, of nickel-steel for piston-rods, crank-pins and other moving parts of engines which are subjected to severe strains. On this subject Mr. J. E. Johnson, of Longdale, Va., in a paper read before the American Society of Mechanical Engineers,[‡] relates the experience of the Longdale Iron Co. with piston-

---

[*] H. K. Landis, *Scientific American*, January 9, 1897.
[†] A. L. Colby, "High-Carbon Steel for Forgings," *Eng. Club of Phila.*, Nov., 1896.
[‡] *Trans. Am. Soc. Mech. Engrs.*, vol. xix.

rods in Baldwin locomotives. Two piston-rods of ordinary machine-steel, after about fourteen months' service, broke and smashed the cylinder-heads. Soft Swedish iron was then tried, and after four months' service one of these broke in the same way. The breaking of the rod of soft material was not very surprising, and was met by ordering material for a set of rods of high-carbon and one of nickel-steel. These have been in use over two years and have given perfect satisfaction.

Railway-Axles.

The rigidity and elasticity which nickel-steel manifests under drop-tests, and which have been examined under the head of the rigidity of nickel-steel, have led to its use in railway locomotive- and car-axles. Over 600 freight-cars of 100,000 pounds capacity, for the Pittsburgh, Bessemer and Lake Erie railway, have been equipped by the Carnegie Steel Co. with axles of nickel-steel. Of the tests of these axles, shown in Table XVII., Mr. Wood says:*

"When carbon-steel breaks under the falling hammer, it breaks short and goes 'all at once.' Nickel-steel shows much greater rigidity than carbon-steel; and when it does break, it yields by a gradual cracking-in from the opposite sides.

"Nickel-steel gives warning; carbon steel does not. Axles of 0.25 carbon and 3 to 3.25 per cent. nickel are as stiff as 0.40 to 0.45 carbon; that is to say, the amount of deflection produced by the blow of the falling weight is about the same in a 0.25-carbon nickel-steel as in a 0.40 to 0.45-carbon simple steel, while the nickel-steel is much tougher and will withstand 50 per cent. more blows."

Railway-Tires.

When a locomotive at full speed strikes a curve, the safety of the train depends upon the resistance of the tire-flanges on the rails to the tangential force, which, if unchecked, would carry the locomotive forward in a straight line. It is customary in England to use low-carbon, and in the United States to use high-carbon steel for railway-tires. In either case the very best material obtainable (that which shows the greatest resistance to strain, combined with the power of resisting a sudden shock) is emphatically necessary.

"The usual specifications for railway-tires," says Mr. Beardmore, " demand that they shall stand compression one-sixth of their diameter without cracking."

---

* Mr. E. F. Wood, Carnegie Steel Co. Private communication.

From a number of nickel-steel tires made at the Parkhead Forge, in Glasgow, one containing 0.18 carbon and 3 per cent. nickel gave the following physical tests:

TABLE XXXII.—*Test of Nickel-Steel Tire.*

Elastic limit,	55,000 lbs. per sq. in.
Ultimate strength,	87,000 " "
Elongation,	28.7 per cent.
Contraction,	46 "

This tire, $39\frac{1}{2}$ inches in diameter, was pressed down to 19 inches (three times what the specifications demanded) without showing signs of fracture. These results could not have been obtained from any other metal of 0.18-carbon known to me.*

Hull-Plates.

The collision or grounding of steel vessels brings an enormous and suddenly-applied force to bear upon a small area, the effect being similar to the blow of an armor-piercing shell. Under such circumstances the brittleness of carbon-steel offers much contrast to the toughness of nickel-steel. Says Commander Eaton:[†]

"A comparative test between a nickel-steel plate and a carbon-steel plate, made with a view to subjecting them to the same strains as those experienced by vessels grounding, disclosed the fact that nickel-steel was far better adapted for plating and frames than ordinary steel. Both plates were riveted to angles in a manner intended to imitate the riveting of a ship's plate between the frames. A round-faced punch, placed on each plate, was then struck by a heavy falling weight. The diameter of the punch was 3 inches, and the head was carefully rounded to preclude cutting edges. Each plate endured 13 blows before rupture; but at the next blow each plate showed a clean aperture; that in the carbon plate was 23.1 square inches in the clear, while the aperture in the nickel plate was but three-quarters of a square inch.

"If these holes had been opened in a ship's bottom, at a depth of 25 feet, the rates of inflow would have been 11 tons per minute in the carbon-plate, and 0.36 ton in the nickel-plate. It would seem that comment on these figures is unnecessary. It should be stated that neither plate was annealed, an omission which counted more heavily against the nickel than against its competitor. The nickel-plate preserved its shape, and in a remarkable degree absorbed the energy of the blows without transmitting wrecking-strains to the angles. The carbon plate was twisted and bulged, and the angles holding it were bent so badly that the whole structure was on the point of being wrecked."

---

* Wm. Beardmore, "Nickel Steel, etc.," *Inst. Naval Arch.*, April, 1897.

† "The Advantage of Nickel-Steel Over Carbon-Steel." See Bibliography at the end of this paper.

The fact that loss by corrosion of nickel-steel in salt and fresh water is much less than carbon-steel* constitutes an additional factor in its fitness for hull-plates.

In nickel-steel for the hull-plates of vessels a tensile strength of 85,000 pounds, with an elastic limit of 60,000 pounds, has been obtained, together with an elongation of 20 per cent. Similar plates of ordinary steel would have a tensile strength but little over the elastic limit of nickel-steel, with an elastic limit about one-half that of nickel-steel.† In the case of steamships, the saving of weight that could be effected by the use of nickel-steel, and the increased capacity for coal or freight thereby attained, would increase greatly the earning power of the invested capital.

The British Admiralty requirements for ships' plates are from 26 to 30 tons (58,240 to 67,200 pounds) tensile strength. Nickel-steel (about 3 per cent. Ni) gives 52 tons (116,480 pounds) tensile strength, and the *yield-point* is 28 to 30 tons (62,720 to 67,200 pounds), equal to the *ultimate* tensile strength required by the Admiralty.‡

Armor-Plate.

The adoption by the United States government of nickel-steel as a material for armor-plate dates from the tests made at Annapolis, in 1890, of simple carbon-steel plates, compound plates of steel and wrought-iron, and unhardened nickel-steel plates. The results of these trials are too well known to need further comment. The resistance obtained with the nickel-steel plate far exceeded the expectations. At the fifth shot, with a striking energy of 4988 foot-tons, an 8-inch armor-piercing projectile broke in many pieces, after having forced its point but 10½ inches beyond the back, and without developing a sign of a crack. Never before in the history of modern armor had a plate given results to compare with these, withstanding a total energy of 16,940 foot-tons,—1835 foot-tons per ton of plate—without developing a single crack, and without being perforated.§

---

* See page 38.

† "Nickel-Steel, etc.," H. A. Wiggin, *Iron and Steel Inst.*, August, 1895.

‡ "On Nickel-Steel," Wm. Beardmore, *Inst. Engrs. and Shipbuilders of Scotland*, March, 1896.

§ "The Year's Naval Progress," July, 1891, *Proc. United States Naval Inst.*, Vol. xxi., No. 5 ; *Jour. Franklin Inst.* Also Report of Secretary of United States

These results proved, beyond question, the superiority of nickel-steel as a material for armor-plate, and led to the decision of the Armor Board to use nickel-steel armor for the new battle-ships and cruisers of the United States navy. The average composition of the steel used is 3.25 per cent. nickel and 0.25 carbon. This forms an exceedingly tough and homogeneous metal, in which hardness is afterwards produced by the Harvey process of carburization. The finished armor-plate contains on the exterior face, about 1 per cent. carbon, which gradually diminishes until, at a depth of 2.5 inches, the original composition of 0.25 carbon is found. This plate can now be oil-tempered, annealed or subjected to any other heat-treatment desired. Admiral Sampson, speaking of the use of nickel in armor-plate, says :*

"To any metallurgist acquainted with the infinitude of results that may be obtained by a variation in the composition and treatment of simple steel, the advantageous possibilities arising from the introduction of so benign an ingredient as nickel must be apparent. In other words, where simple steel is strong and tough, both qualities may be improved by adding the proper amount of nickel. The susceptibility of nickel-steel to treatment is remarkable, and yet this steel may be abused in the most shameful way without failure. Nickel appears to render the carbon more sensitive to hardening, and hence water-hardened Harvey plates of nickel-steel are toughened at depths hardly affected in simple steel plates."

The adoption of nickel-steel as material for armor-plate by the United States government has led to its use by all the great powers; and at the present time nickel-steel is used, either in part or exclusively, as material for armor-plate and protective deck-coating in nearly all modern war-vessels, wherever constructed.

Structural Beams and Shapes.

Mr. H. H. Campbell† observes that the strength of a bridge of a given weight and design, the weight of a bridge of a given strength and design, and the longest span possible under a given design are all limited by the elastic strength of the mate-

---

Navy; and F. L. Garrison, "Development of American Armor-Plate," *Jour. Franklin Inst.*, June, 1892.

* "The Present Status of Face-Hardened Armor-Plate," *Soc. Nav. Arch. and Mar. Eng.*, 1894.

† *Am. Soc. Civ. Engrs.*, June, 1895.

rial used. These well-understood axioms point naturally to the use of steel possessing high elastic limit; but this tendency is opposed by the no less well-known fact that as the elastic strength of steel is increased by ordinary strengthening-influences, there is a reduction either in the static or the shock-ductility.

Mr. R. W. Davenport* remarks that an examination of the physical characteristics of this metal (nickel-steel) shows it to possess valuable qualities which explain its toughness and resistance to shock. The effect of nickel upon the elastic limit is, however, of the greatest importance, as it raises this quality in a marked degree, relative to the tensile strength, and thus insures a combination of elastic strength and ductility, or toughness, unknown in any other metal.

Nickel-steel shapes, beams, and T and Z bars, made for the United States navy, have been tested by Commander J. G. Eaton.

"The chemical and physical qualities are typified in the following record :

"Carbon, 0.25 ; sulphur, 0.023 ; manganese, 0.68 ; phosphorus, 0.010 ; nickel, 3.30 ; tensile strength, 87,580 pounds ; elastic limit, 62,700 pounds; elongation in 8 inches, 22 per cent. ; reduction of area, 52 per cent.

"The Z bar was one of a lot and not specially selected. It was unannealed and of basic open-hearth. Specimens 2 inches in width and 12 inches long closed cold on themselves, without sign of fracture and with a total absence of hair-cracks. The same bar was then annealed at a temperature of 1600° F. and again tested as follows :

"Tensile, 84,640 pounds ; elastic limit, 59,500 pounds; elongation in 8 inches, 24.5 per cent. ; reduction of area, 54 per cent.

"A bar of mild steel of the same chemical composition, but without the nickel, showed on similar tests :

"Tensile, 62,410 pounds ; elastic limit, 36,420 pounds; elongation in 8 inches, 26.25 per cent. ; reduction of area, 58.5 per cent.

"It needs only a glance at these figures to note that the gain to the working-load has been increased in the ratio of 36 to 59. or practically 64 per cent., while the elongation is at a figure sufficient for all ordinary purposes."†

The increase in elastic limit by the use of nickel in steel makes possible an increase in the span, or a decrease in the weight, of bridge-sections. The resistance of nickel-steel to jarring shock renders its use in eye-bars for bridge-construction particularly desirable. Such eye-bars are subject to repeated jarring strains; and the failure of an eye-bar means the collapse

---

* *Cassier's Mag.*, August 18, 1897.

† Commander J. G. Eaton, U. S. N., "The Advantages of Nickel-Steel."

of a bridge. The necessity thus arising for the use of a steel with high elastic limit (about 70,000 pounds) has led to the use, in some cases, of 0.40-carbon simple steel. This percentage of carbon renders steel sensitive to heat-treatment; and, in heating to forge the eye, the bars are in danger of becoming crystalline by over-heating. The same high elastic limit can be obtained by using an 0.18- to 0.20-carbon steel, with 3 to 3.5 per cent. of nickel, which steel is much less liable to damage by overheating than the simple steel of higher carbon.*

Rivets.

Mr. Beardmore† reports that, having satisfied himself that nickel-steel was reliable, as regards punching, he had some rivets made of this metal, which were found to be considerably tougher than ordinary rivets. They gave a tensile strength of 35.7 tons (79,968 pounds) per square inch, with an elongation of 36 per cent. in 1.25 inches, and a contraction of area of 60 per cent., and were very ductile. An ordinary and a nickel-steel rivet were nicked with a chisel and hand-hammer and then broken. In the case of the nickel-steel the fracture was fibrous and the metal appeared to have torn gradually, whilst the ordinary carbon-steel rivet had broken short off.

Some very interesting and important experiments were recently made at the works of the Bethlehem Steel Co. by the Engineer of Tests, Mr. Maunsel White, for the purpose, not only of ascertaining the reliability and comparative efficiency of nickel-steel for general riveting, but also of observing the effects of working rivets of this material at different degrees of heat. The results were such as to remove all doubt as to the possibility of working these rivets safely within perfectly reasonable limits and with the exercise of only ordinary care. Two series of tests were made, the arrangement of the plates being for the first test as shown in Figs. 4 and 5, and for the second test as shown in Fig. 6.

The results of the first series of tests were as follows (see Fig. 7):

---

* E. F. Wood, Carnegie Steel Co.   Private communication.

† "Nickel-Steel," *Inst. Engrs. and Shipbuilders of Scotland*, March, 1896; *Industries and Iron*, May 1, 1896.

Fig. 4.

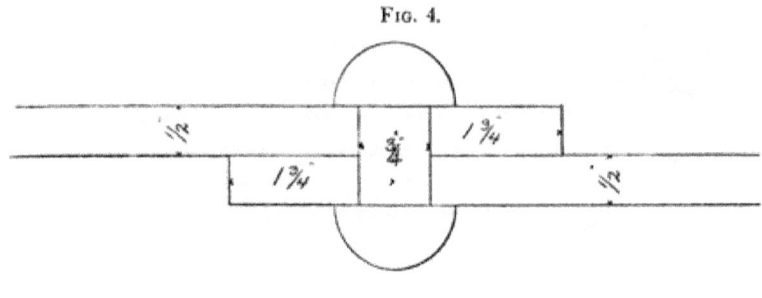

Fig. 5.

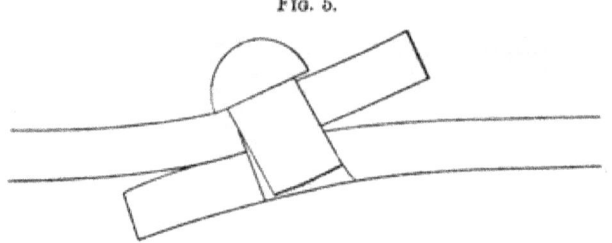

Method of Testing Nickel-Steel Rivets—Single Shear.

Fig. 6.

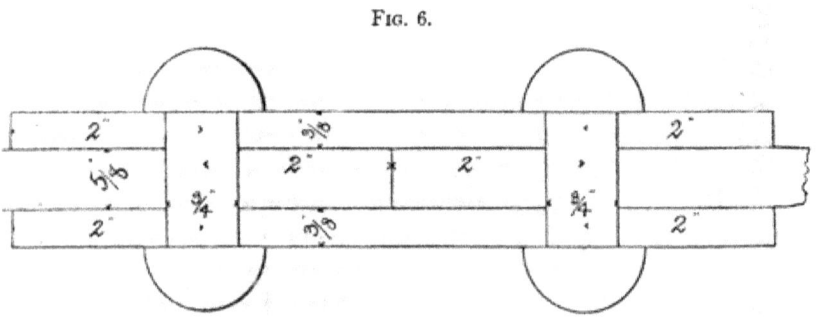

Method of Testing Nickel-Steel Rivets—Double Shear.
(From *Jour. Am. Soc. Naval Engrs.*)

NICKEL-STEEL; A SYNOPSIS OF EXPERIMENT AND OPINION. 57

Fig. 7.

Nickel-Steel Rivets—Single-Shear Tests. (See Table XXXIII.)
(From *Jour. Am. Soc. Naval Engrs.*)

Fig. 8.

Nickel-Steel Rivets—Double-Shear Tests. (See Table XXXIV.)
(From *Jour. Am. Soc. Naval Engrs.*)

NICKEL-STEEL; A SYNOPSIS OF EXPERIMENT AND OPINION. 59

FIG. 9.

CENTER LINE OF FIRST ROD AFTER BENDING	LENGTH FEET	DEFLECTION FEET	CENTER LINE OF SECOND ROD AFTER BENDING	LENGTH FEET	DEFLECTION FEET	PRESENT NICKEL STEEL ROD
				41	.000	
	40	.000		40	.000	
	39	.000		39	.000	
	38	.000		38	.000	
	37	.000		37	.005	
	36	.000		36	.010	
	35	.000		35	.016	
	34	.000		34	.024	
	33	.000		33	.034	
	32	.000		32	.051	
	31	.000		31	.036	
	30	.005		30	.042	
	29	.008		29	.047	
	28	.010		28	.052	
	27	.012		27	.057	
	26	.016		26	.062	
	25	.020		25	.072	
	24	.021		24	.078	
	23	.025		23	.083	
	22	.028		22	.094	
	21	.031		21	.104	
	20	.042		20	.130	
	19	.052		19	.208	
	18	.063		18	.245	
	17	.073		17	.312	
	16	.089		16	.396	
	15	.104		15	.464	
	14	.125		14	.521	
	13	.135		13	.588	
	12	.145		12	.609	
	11	.161		11	.615	
	10	.162		10	.599	
	9	.167		9	.562	
	8	.177		8	.500	
	7	.167		7	.432	
	6	.156		6	.354	
	5	.130		5	.292	
	4	.104		4	.208	
	3	.078		3		
	2	.003		2	.047	
	1	.000		1	.000	

Vertical Scale 3⁄4 = 1 Foot
Horizontal Scale 3⁄4 = 1 Foot

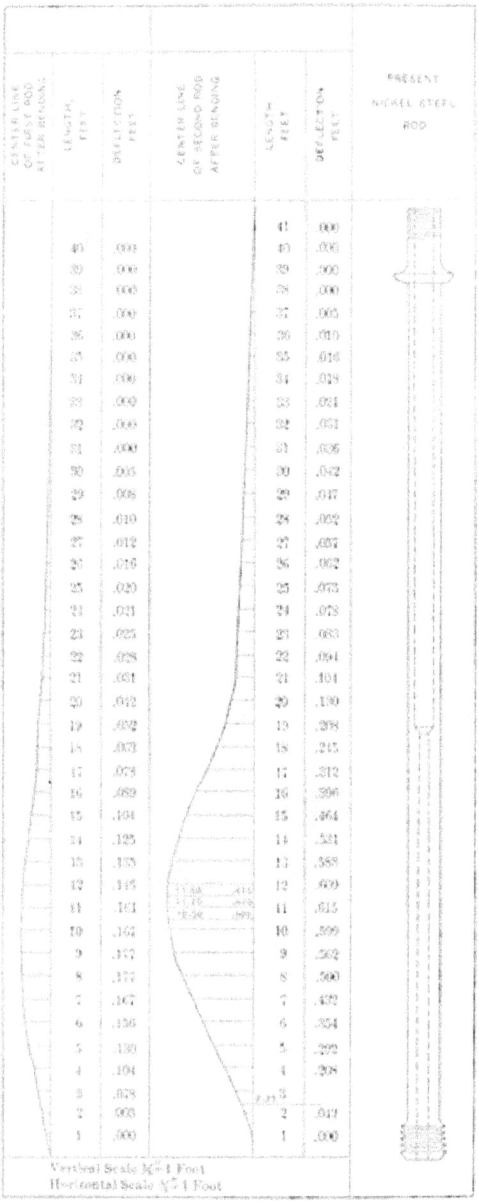

COMPARATIVE FAILURES OF HIGH AND LOW-CARBON STEEL PISTON-RODS OF 125 TON HAMMER, BETHLEHEM IRON CO., BY H. K. LANDIS.

(From *Jour. Franklin Inst.*)

FIG. 10.

Stages in the Manufacture of Nickel-Steel Tubes, by the Pope Manufacturing Co., Hartford, Conn.
(From *Iron Age*.)

Fig. 11.

Making Nickel-Steel Tubes.

Fig. 12.

Straightening Nickel-Steel Tubes.
(From *Iron Age*.)

Fig. 13.

Nickel-Steel Slabs and Bicycle-Tubing, at the Works of the Pope Manufacturing Co., Hartford, Conn.
(From *Iron Age*.)

Fig. 14.

Interior of Works of the Crescent Steel Co., Pittsburgh, Pa., Manufacturers of Crucible Nickel-Steel.

TABLE XXXIII.—*Tests of Nickel-Steel Rivets.*

¾-INCH NICKEL-STEEL RIVETS—SINGLE SHEAR.

Mark.	Breaking-Load. Lbs.	Nature of Fracture.	Shearing-Load. Lbs. per Sq. In.
1 A	84,800	Plate broke.	88,330
1 B	87,600	Rivets sheared.	91,240
1 C	80,700	" "	84,060
1 D	78,400	" "	81,660
2 A	70,600	Heads broke off.	73,530
2 B	85,700	Plate broke.	89,270
2 C	90,900	Heads broke off.	94,680
2 D	79,700	" "	83,000

The letters A, B, C and D refer to the heat used in riveting, as follows: A, bright cherry-red; B, light-red; C, yellow; D, almost white. The numerals 1 and 2 coupled with these letters refer to the grades of nickel-steel used in the rivets.

The second series of tests was made with rivets in double shear. In these tests, two rivets in each plate were driven at different heats. In four joints, heats A and B were used, and in the other four joints heats C and D. The results of this second series, double-shear tests, were as follows (see Fig. 8):

TABLE XXXIV.—*Tests of Nickel-Steel Rivets.*

¾-INCH NICKEL-STEEL RIVETS—DOUBLE SHEAR.

Mark.	Breaking-Load. Lbs.	Nature of Fracture.	Shearing-Load in Rivets. Lbs. per Sq. In.
1 AB	78,550	Plate sheared.	87,200
1 AB	81,600	Rivet broke on one side.	90,550
1 CD	83,800	Plate sheared.	93,100
1 CD	74,050	Rivet sheared	82,270
2 AB	84,200	Plate sheared.	93,550
2 AB	82,640	" "	91,820
2 CD	79,800	" "	88,660
2 CD	84,100	" "	93,440

In order to make some comparison between nickel-steel and the ordinary steel rivets, two test-pieces were made with ¾-inch ordinary steel rivets, and were marked E and F. In single-shear, ½-inch steel plate was used, with holes 2 inches from edge of plate. In double-shear, ⅜-inch steel plates were used, with ⅜-inch steel-plate welts, and the holes were 2 inches from the edge. The results were:

TABLE XXXV.—*Tests of ⅜-Inch Common Steel Rivets.*

Mark.	Breaking-Load. Pounds.	Nature of Fracture.	Shearing-Load in Rivets. Pounds per Square Inch.
E,	53,200	Rivets sheared.	43,600 single shear.
F,	55,100	" "	46,000 double "

From these results Mr. White says it may be safely deduced that a ¾-inch nickel-steel rivet will replace a $1\tfrac{1}{16}$-inch, or even possibly a 1⅛-inch, common steel rivet, thus effecting a saving of considerable plate-section and giving increased strength.*

These results, obtained in practical work, offer a very strong corroboration of the results deduced by Prof. Rudeloff in his experiments on the effect of nickel upon the shearing-strength of iron, referred to on page 29.

Steam-Hammer and Rock-Drill Piston-Rods.

The piston-rods for steam-hammers are particularly liable to what has been termed molecular fatigue: they are subject to continued shock, to resist which they must possess a very high elastic limit.

Fig. 9 shows the experience of the Bethlehem Steel Co. with rods for their 125-ton hammer.† The first rod was made of high-carbon open-hearth steel, hydraulically forged, with a 4-inch hole running through the entire length, and having an outside diameter of 16 inches. The failure of this rod was attributed to partial crystallization and to its being too high in carbon, and, therefore, too brittle. These were the usual reasons given for failure under similar conditions in those days. Consequently, the next rod was made of a steel much lower in carbon, in the hope that it would withstand the effects of shock better. This supposition proved to be erroneous, and this rod failed after a shorter service than the first, and its failure was much more serious. About this time the advantages of material with high elastic limit became known, and the third rod was made of nickel-steel. The dimensions were changed somewhat: the outside diameter was changed from 16 to 17 inches, and the hole was made 8 inches in diameter for 22 inches from

---

* "Nickel-Steel Rivets," *Jour. Amer. Soc. Naval Engrs.*, November, 1898. Figs. 4 to 8 inclusive have been taken from this paper.

† This statement is taken from the paper of Mr. H. F. J. Porter, *Jour. Frank. Inst.*, December, 1897.

the top; the balance was made 7 inches instead of 4 inches, as before. This rod has fulfilled all expectations. The amount of deflection of the two rods which failed is shown in the figure, which exhibits also the character of rod now made for users of rods who want something that will last a long time.

At the Crescent Steel Works, in Pittsburgh, nickel-steel rods were substituted for carbon-steel in a battery of steam-hammers over a year ago. The hammers range from 300 pounds to 7 tons, the same material (3 per cent. nickel and about 0.30 carbon) being used in all. These rods have thus far given perfect satisfaction.*

In oil-tempering steel-forgings the objects to be tempered are suspended from a bar which carries a spider, and this bar is plunged into the oil-bath together with the object to be tempered. A bar of rectangular section subjected to this repeated heating and cooling tends to become cylindrical in section, and, if made of simple steel, will develop a longitudinal crack. As nickel-steel will, in this position, last much longer than simple steel, it is used for this purpose in the Bethlehem Steel Company's tempering- and annealing-department.

The same quality which enables nickel-steel to resist rapidly alternating strains renders it very serviceable in rock-drills, for which purpose it is largely used. The Sullivan Machine Company says on this subject:

"Some years ago we made some experiments in the use of nickel-steel in parts of machines where we had previously used iron or open-hearth or crucible-steel. The parts mentioned were particularly liable to breakage, either from strain or on account of crystallization. After an experience of several years, during which we have considerably extended the use of the nickel-steel, we have become satisfied that it is an improvement over the iron and steel we had been using (though we had always used the best of ordinary products), the machine parts made from nickel-steel being less liable to break on account of weakness or crystallization. We believe also that the nickel renders the wearing parts of machines, which run on other parts, less liable to cut—the nickel apparently having the property of making the surfaces smooth with wear, even though not always properly oiled."†

This statement concerning the reduction of friction by the use of nickel is corroborated by the remarks of Mr. H. T. Williams, quoted above, under "Influence of Nickel on Hardness," and also below, under "Nickel in Tool-Steels."

---

* Mr. McDonald, Crescent Steel Works. Private communication.
† T. W. Fry, Sec. Sullivan Machine Co., Claremont, N. H. Private communication.

Bicycle-Tubing.

On this subject Mr. Harold E. Eames* says, that in approaching the consideration of the selection of a suitable material for bicycle-tubes, it should be borne in mind that not only must the bicycle provide strength enough to resist the maximum shock of any kind likely to be incurred in ordinary riding, but we must also insure that the life of the structural parts of the machine under the same conditions equals, if possible, that of the bearing and working parts. He adds that it would be necessary merely to select that composition of steel or its alloys which gave the greatest strength in the annealed state, and that steel alloyed with about 5 per cent. of nickel seems to fill this first requirement better than any other material in what might be called commercial existence.

This tubing, containing 5 per cent. of nickel, is made and largely used by the Pope Manufacturing Company, of Hartford, Conn. In its circular on the use of tubing the Co. says:†

"For handle-bars and shafts, seat-posts and rods, which admit of oil-hardening after all the brazing operations have been completed, nothing can approach the 5 per cent. nickel in rigidity and safety. In the oil-hardened state this material has developed an ultimate strength of not far from 240,000 pounds per square inch, with an elongation ample to afford insurance against sudden rupture. An oil-hardened nickel-steel handle-bar is both strong and tough, and fortunately is altogether beyond the realm of experiment, as not less than 150,000 of them have been produced and tested in actual use, with the greatest satisfaction to the manufacturer and rider. The use of this material also solves the difficult question of providing a seat rod which will really resist the collapsing effect of the saddle-clamp and a seat-post which will resist the effect of the head-clamp."

The following is a description of the manufacture:

From the plate of nickel-steel received from the manufacturer a circular disc is stamped out, as shown at the left in Fig. 10, which illustrates the several operations of cupping to which this steel is subjected. This blank is then taken to the double-acting hydraulic press; the blank is held between the two surfaces of the outer slide, while the center or drawing-slide, which has a much longer stroke, pulls the metal from between these surfaces and forces it through the center die, the result being a

---

* "Consideration of Material for Bicycle Tubing," *N. Y. Iron Age*, June 18, 1896.

† Pope Mfg. Co. Circular on "Pioneer Tubing."

comparatively wide and shallow cup. The cup thus formed is absolutely free from any indication of wrinkle or crack at the side; and there is no evidence that the metal has been in any way tortured. The cup at the second drawing is reduced in diameter and considerably lengthened, the final operation being done upon the hydraulic draw-press, which is single-acting, *i.e.*, it is provided with only one plunger, which has a stroke long enough to give the shape shown in Fig. 10. The completed tube is shown in the same engraving, and one of the 0.50-carbon steel billets.* Figs. 11, 12 and 13 (taken, together with Fig. 10, from the *Iron Age*) further illustrates this manufacture.

Nickel in Tool-Steels.

The advantages gained by the use of nickel in the soft open-hearth steels seem to warrant the expectation that a corresponding gain would accrue from its use in high-carbon crucible-steels. The difficulties attendant upon the formation of such an alloy have deterred many from using nickel in tool-steels;—the ease with which nickel absorbs oxygen introducing an element of uncertainty into its manufacture.

Mr. H. J. Williams, of the Damascus Steel Co., of Pittsburgh, has given to this subject much experimental investigation, and the results of his researches are here for the first time made public. Mr. Williams says:

"The difficulty in the manufacture of nickel-steel for tools is caused by the tendency of nickel in the presence of high-carbon to develop seams, unsoundness and partial oxidation. These difficulties have been overcome, and sound billets of nickel-steel of any desired percentage in nickel or carbon can now be obtained.

"The working of these billets into tools or wire-rods presents no especial difficulties. The nickel-steel works tougher than simple steel; causes more rebound of the hammer and necessitates a lighter draught on the rolls or dies.

"For tools, the alloy containing about 0.80 carbon, with 3 per cent. of nickel, is probably the best alloy for every-day use. Increasing the nickel to 5 or 6 per cent. increases the strength,

---

‡ "The Manufacture of Bicycle Tubing; a Description of the Pope Tube Works," *Iron Age*, Jan. 7, Jan. 14 and Mar. 4, 1897.

but necessitates skilled treatment in forging. Tool-steel with 0.80 carbon and from 3 to 5 per cent. nickel, possesses all the hardness of high-carbon simple steel, without its brittleness. Nickel-steel shows very pronounced toughness. If nicked when hot, and then struck on the anvil after it has cooled, it is very liable to break under the hammer—not at, but beyond the nick; while carbon-steel almost invariably breaks at the point where it was nicked when hot. This nickel tool-steel has a satin grain and does not crack on quenching as readily as simple steel; the tensile strength runs as high as 260,000 pounds per square inch. A nickel-steel tool containing 4 or 5 per cent. of nickel, with 0.80 carbon, is as hard after tempering as the best simple tool-steel of 1.00 to 1.25 carbon, without the brittleness and glassy nature which distinguish the latter. When used for drills, the nickel-steel appears to work with much less than the usual friction, and does not become so hot in use as carbon-steel.

"Up to 6 per cent. of nickel, the hardness obtained by quenching is as largely dependent on the amount of carbon present as in simple steels, the only difference being that nickel-steels appear to attain a desired temper with 0.20 to 0.30 per cent. less carbon than simple steels. Nickel-steels are consequently exceedingly sensitive to heat-treatment, but, on the other hand, are less easily spoiled by overheating, and therefore may be worked at somewhat higher heat than a carbon-steel that will give the same temper."*

If the nickel be increased to over 6 per cent., tempering and hardening by quenching does not seem to have much effect upon the metal. This is especially noticeable when nickel rises above 12 per cent.; and at 18 per cent. the influence of carbon is almost neutralized. In the 18 per cent. nickel-steel the same hardness is obtained with 0.30 carbon as with 0.75 carbon, and the metal possesses its maximum of strength and elasticity at about this percentage of nickel. Further increase in nickel causes an openness of structure and a consequent softening.†

---

* H. J. Williams, Damascus Steel Co. Private communication.

† See Rudeloff, *Verein zur Beförderung des Gewerbfleisses*, Feb., 1896, p. 82. "Elasticity under pressure increases with increase in nickel, and the change of form by pressure decreases with increase in nickel up to 16 per cent."

Steel with 18 per cent. nickel and 0.60 carbon is perceptibly softened by heating and quenching; but if heated to redness and forged on the anvil till cold, it attains the same elasticity as a higher-carbon steel quenched in the usual manner. The 18 per cent. nickel-steel can readily be worked hot, but is exceedingly hard to punch and press cold.*

Steel with 25 per cent. nickel is very much softened by heating and quenching, and can be worked cold almost as readily as German silver.*

As the 18 per cent. nickel-steel is silvery white in color and has a soft luster, more resembling silver than steel, this alloy is particularly useful for table-ware and cutlery. For this purpose its non-corrodibility gives it a great advantage over simple steel, and its temper and hardness make it superior to German silver.

An alloy of 3 per cent. nickel-steel with tungsten, and containing over 2 per cent. carbon, forms a self-hardening steel, similar to Mushet steel, with the advantage that it can be forged at the same heat as ordinary tool-steel.*

Hydraulic Cylinders.

The enormous power applied in the hydraulic compression of armor-plate and heavy ingots demands the use of exceptionally strong material in the cylinders. Nickel-steel has been very successfully used for this purpose at some of the largest forges in this country and Great Britain.

At the Parkhead forge, in Glasgow, a cylinder for a 12,000-ton hydraulic press was cast of nickel-steel. It weighed, as cast, 143,360 pounds, and was 72 inches in diameter. The machined cylinder weighed 76,160 pounds and showed an elastic limit of 56,000 pounds, with a tensile strength of 92,200 pounds per square inch.†

At the Carnegie Works, at Homestead, Pa., it was discovered in 1896 that the five-piece cylinder of the 10,000-ton forging-press was developing a crack. As the makers could not replace this cylinder in less than eleven months, it was hastily decided to cast a cylinder of open-hearth steel, using enough

---

* H. J. Williams, **Damascus** Steel Co. Private communication.
† Wm. Beardmore, "Nickel-Steel," *Inst. Naval. Arch.*, April, 1897.

nickel to insure maximum strength. A hollow ingot, of 0.25 carbon, with 6 per cent. nickel, weighing with the sinkhead 140 tons, was successfully cast. This is believed to be the largest hollow casting of nickel-steel yet produced. Considerable difficulty was experienced in erecting machinery of sufficient size to handle such a casting, but an impromptu lathe was finally erected, and the ingot was turned to the finished shape, in which it weighed 90 tons. This cast cylinder has now been in use for nearly two years and shows no signs of wear. It withstands an interior pressure of three gross tons to the square inch, and is undoubtedly the largest casting yet made of nickel-steel for hydraulic cylinders.*

Rifles and Small Arms.

The specifications for the Lee straight-pull rifles, supplied to the U. S. Government, read in part as follows:

"The material of which the barrels are made must be forged or rolled steel, oil-tempered and then annealed, and showing, on 2-inch specimens of standard form, an elastic limit of at least 80,000 pounds per square inch, and an elongation of at least 20 per cent. The above requirements can be met by an open-hearth steel containing about 4.5 per cent. of nickel; and such steel will be preferred."

The high ratio of elastic limit to ultimate strength conferred upon steel by the addition of nickel constitutes its chief merit for this purpose. No matter what the arm be, from revolver to hundred-ton gun, it is very desirable that the strength thereof should be *available* strength. In simple steel, annealed and tempered, the elastic limit of the metal is about 60 to 70 per cent. of the ultimate; that is to say, the gun-barrel may be deformed by a strain not much over half that required to produce rupture. For accuracy and safety it may be said that deformation is fully as undesirable as destruction of the piece. Any means whereby the amount of available strength in the weapon can be increased, and the useless strength and dead weight thereof can be decreased, is a boon to the manufacturer of arms and ordnance.

The Société Cockerill at Seraing, in Belgium, requiring for artillery and munitions of war a metal of very high elastic limit, and with sufficient malleability and ductility to insure its

† E. F. Wood, Carnegie Steel Co. Private communication.

working with ease, their engineer, Mr. Moulan, began five or six years ago to experiment with alloys of nickel and iron. Among other alloys, one of 7.5 per cent. nickel in pure homogeneous iron was produced. This metal seemed so nearly to fulfill the requirements that complete tests were made for comparison, both with iron and hard steel.

TABLE XXXVI.—*Tests of Ferro-Nickel, Iron and Steel.*

(Recalculated by H. K. Landis.)

Metal.	Limit of Proportionality. Lbs. per Sq. In.	Elastic Limit. Lbs. per Sq. In.	Ultimate Strength. Lbs. per Sq. In.	Elongation. Per cent.	Reduction of Area. Per cent.
Ferro-nickel.					
Not tempered	35,270	70,389	76,800	24.3	60.4
Tempered at 900° C. in water	64,300	142,000	178,500	10.2	50.5
Ditto, and annealed at 500° C.	59,450	117,000	118,000	12.5	61.2
Tempered at 900° C in oil	55,750	139,000	142,200	9.3	42.3
Ditto, and annealed at 500° C.	49,790	115,800	120,000	12.2	52.5
Homogeneous Iron.					
Not tempered	16,500	29,870	53,900	29.4	64.9
Tempered at 900° C. in water	25,600	46,900	69,100	23.4	57.4
Ditto, and annealed at 500° C.	16,780	39,000	70,500	34.6	67.9
Tempered at 900° C. in oil	22,200	44,950	62,000	29.4	66.2
Ditto, and annealed at 500° C.	20,760	34,150	54,200	29.2	67.7
Hard Steel (0.55 C.).					
Not tempered		73,400	122,000	12.1	24.4
Tempered at 900° C. in water		75,600	105,000	22.	.9
Ditto, and annealed at 500° C.		114,000	143,500	7.7	27.3
Tempered at 900° C. in oil		101,800	132,800	1.8	4.7
Ditto, and annealed at 500° C.		112,000	150,760	9.8	27.3

As will be seen from these tests, all the advantages are on the side of ferro-nickel. This alloy, after oil-tempering and annealing, shows an elastic limit of 90 per cent. of the ultimate strength; while in the hard steel the same treatment raised the elastic limit only to 74 per cent. of the ultimate. In practice this means that, of the total weight of a gun-barrel, 36 per cent. is, in the case of carbon-steel, useless weight; while in ferro-nickel only 10 per cent. of the weight is unavailable in action. Since these tests were made, Mr. Moulan has used ferro-nickel alone for field-pieces and artillery-equipment.*

---

* *Stahl und Eisen*, 1895, vii., p. 346; *Rev. Univ. des Mines*, xxvii., p. 152; *Industries and Iron*, Nov. 2, 1894.

Resistance-Wire.

The high electrical resistance of alloys of iron with 25 to 30 per cent. of nickel has led to its use in electrical machines instead of German silver. "Tico" resistance-wire, made by the Trenton Iron Company and containing about 28 per cent. of nickel, is largely used for resistance-coils in rheostats and electrical heaters. It has considerable advantage over German silver in having three times as much resistance, and retaining its strength and elasticity even after heating to bright redness. No. 14 B. & S. Tico wire averages 8 feet to the ohm of resistance; the same size of Geman-silver wire requires in practice 25 feet to the ohm.*

A similar resistance-wire, known as "Climax" wire, containing about 24 per cent. of nickel, is made by Roebling's Sons Co. This wire has a specific gravity of 8.137, a resistance per mill-foot at 75° of 504.86 ohms, and a temperature-coefficient of 0.042 per cent. per degree F.

Another nickel-iron resistance-wire, made in Germany, and containing about 29 per cent. of nickel, has specific gravity 8.4; specific resistance at 20° C. (68° F.) of 86 microhms; a coefficient of temperature of 0.00065 for 1° C., and a resistance per mill-foot of 517.5 ohms.

The specific resistance of German silver is on the average, at 20° Centigrade, 31.5 microhms.

Owing to their high percentage of nickel, these wires are nearly incorrodible, and, owing to their high tensile strength, they are much less liable to break than German silver. After repeated heating and cooling, German silver becomes brittle, while nickel-iron alloys seem to retain their original elasticity.

Miscellaneous Uses.

There are a number of purposes in which nickel-steel is used, but concerning which commercial rivalry prevents the publication of reliable data. For locomotive fire-boxes, boiler-braces and stay-bolts, bicycle-spokes and chains, torpedoes and torpedo-nets; incandescent lamp-mantle hangers; rolls for tubing and for planished sheets; boiler-tubing, revolvers and small-arms;

---

† J. E. Storey, Storey Motor and Tool Co., Trenton, N. J. Private communication.

safety-deposit vaults; bobbin-spindles; teeth for cotton-pickers, and for a great variety of purposes for which some peculiarity, either of strength, ductility or incorrodibility, gives it especial fitness, nickel-steel is rapidly making its way into popular favor.

The peculiar electrical and thermal coefficients possessed by the alloys of steel with high percentages of nickel, suggest numerous uses in electrical appliances and instruments of precision. All the nickel-steel alloys are remarkably homogeneous, easily worked and susceptible of a high polish.

M. Zetter, director of the French Electrical Equipment Co. (*Compagnie Française d'Appareillage Électrique*), has made numerous experiments on the nickel-steels which lose their magnetism on heating and recover it on cooling. He has used these metals in the manufacture of circuit-breakers, automatic fire-alarms and other instruments, in which an electric circuit is broken by a rise of temperature, produced either by the heating effect of the current or by an external source of heat. He finds these instruments to act with regularity and accuracy.*

M. Dupriez suggests that as the temperature at which magnetism is lost and regained in nickel-steel is a regular function of the amount of nickel present, the 30 per cent. alloy which loses magnetism at 100° C. and regains it 50° may be utilized in the construction of a thermo-electric machine for the direct conversion of heat into electrical energy.†

From Table XXXVII, showing the effect of varying amounts of nickel on the coefficient of expansion, it will be seen that the coefficient falls from 28 to 36 per cent., and then rises again. At 36 per cent. the expansion is only one-tenth that of platinum. Platinum expands about 1 part in 208,800 for a rise of one deg. F. The 36 per cent. nickel-alloy expands 1 part in 2,049,000 parts.

Expressed in the usual way, temperature being read in degrees Centigrade, the expansion-coefficients of some of the common metals are:

---

* Guillaume, "Les Aciers au Nickel," *Comptes Rendus*, March, 1898.
† M. Marcel Dupriez, *Rev. Gen. des Sciences*, Feb. 15, 1898.

TABLE XXXVII.—*Coefficients of Expansion.*

Brass,	0.00001878
Copper,	0.00001718
Soft steel,	0.00001078
Hard steel,	0.00001239
Nickel,	0.00001252
26 per cent. nickel-steel,	0.00001312
28 " "	0.00001131
28.7 " "	0.00001041
30.4 " "	0.00000458
31.4 " "	0.00000340
34.6 " "	0.00000137
35.6 " "	0.00000087
37.3 " "	0.00000356
39.4 " "	0.00000537
44.4 " "	0.00000856
Platinum,	0.00000884
Glass,	0.00000861

From this table it will be seen that a nickel-steel alloy may be made to have a coefficient of expansion which may vary within wide limits. Alloys of the same expansion as glass may be prepared, either with about 29 per cent. nickel or with about 45 per cent. nickel. For all lens-mounts, telescopic and microscopic, and in general for all optical purposes where glass and metal are brought in contact, these alloys have a wide field of usefulness.

For instruments of precision in which extreme accuracy is desirable, the 36 per cent. nickel-steel may be used to great advantage, instead of brass. It expands and contracts with changes of temperature only one-twentieth as much as brass; is very much stronger and more rigid, and hence allows a reduction of weight in the instrument; and, as it is very much less affected by moisture and corrosive gases, and retains a silvery color and polish under adverse circumstances, its advantage for scientific instruments is very evident.

In mercury-pendulums or compensation-pendulums, the weight of mercury or the length of the compensation-bars can be greatly reduced by the use of the 36 per cent. nickel-steel; and by replacing a common pendulum with one of this alloy, the daily variation due to temperature will be diminished in the proportion of 12 to 1.

The incorrodibility of nickel-steel points out its usefulness in pumps, mine-cables, wire-ropes used near roasting-furnaces, and all other situations where necessary exposure to corrosive liquids or gases renders simple steel unserviceable.

V.—Cost of Nickel-Steels.

At the present cost of nickel—33 to 36 cents per pound—each percentage of nickel adds one-third of a cent to the cost per pound of steel. The usual addition of 3 per cent. nickel increases the cost of the raw metal 1 cent per pound. After the steel is cast, the alloy costs no more to work than simple steel, excepting the slight additional cost entailed by keeping nickel-steel scrap in a separate stock-pile.

Finished forgings for moving parts of engines vary in price according to the intricacy of the pattern; and in simple steel forgings the cost may run from 3 or 4 cents per pound up to, and even over, 50 cents per pound of finished product. The use of nickel-steel for fine shafting, connecting-rods, crank- and cross-head pins, may increase the cost 15 to 25 per cent. over the figures for simple steel forgings. The increase in efficiency varies with the amount of carbon present, and runs from 40 per cent. in the soft steels to over 60 per cent. in the hard steels; while in the case of very hard tool-steels, the effect is said to be much greater.

As this gain in efficiency may be utilized either to decrease the weight of the part, or to increase the amount of work done, the net saving effected by the use of nickel is very evident.

Capt. W. H. Jaques, U. S. N., writing in 1895, estimated that even at the then comparatively high price of nickel, there would be a saving of $64,000 in the item of armor alone for each of the British battleships, if nickel-steel carburized armor were substituted for the plain steel Harveyized armor-plate which it was proposed to use.

For hull-plates of large steamships, the use of nickel-steel would effect a large saving in weight and displacement, and the increased freight- or coal-capacity gained thereby would add a large percentage to the efficiency and earning-power of the vessel.

VI.—Manufacture of Nickel-Steels.

Nickel melts at about 1650° C., a heat a little above that of molten steel. The nickel-steel alloys melt, however, at the usual temperatures. Nickel cannot therefore be added in the ladle, as may be done with spiegel or aluminum, but must be alloyed directly in the crucible or open hearth. It may be used either as oxide or as metal. The dry oxide, when pure, contains about 77 per cent. of metal. The usual method of adding nickel, as oxide, in the open hearth, is to enclose the oxide in a rough box of wood or iron, with sufficient charcoal to reduce it to metal. This is placed upon the hearth, scrap- and pig-iron are added, and the charge is worked in the usual way. The use of metallic nickel is, however, to be preferred, as the loss by slagging is not so great. In open-hearth work the loss is estimated as the same as that of iron, viz., 7 to 8 per cent. of the charge.

In crucible-work metallic nickel alone should be used; and the loss in this case is very little. The nickel is added in the crucibles with the rest of the charge.

In working nickel-steels care should be taken to avoid chilling the metal. It is extremely sensitive to sudden, even though slight, changes of temperature and careless handling. Contact with wet or cold ground, exposure to rain, or any other sudden chill may cause a surface-hardening that will interfere with proper working.

In hammering, rolling, pressing or wire-drawing, less draught should be given than with simple steels of the same carbon. In general, soft steels containing nickel should be treated like high-carbon steel. Once the metal is cold, it can be depended upon to resist rough handling and unskilled treatment, but while hot it should be treated with the care and respect due to a metal of such unique properties and phenomenal possibilities.

VII.—Makers in the United States.

The Carnegie Steel Co., the Bethlehem Steel Co. and the Midvale Steel Co. manufacture large amounts of open-hearth nickel-steel for the United States government and for private trade. These firms possess large experience, and every mechanical facility for the manufacture of nickel-steel in any quantity or shape desired.

The Carbon Steel Co., of Pittsburgh, makes a specialty of open-hearth nickel-steel for engine-forgings, locomotive-parts, boiler-sheets and stays, and fire-boxes.

The Crescent Steel Co., of Pittsburgh, deals more exclusively with crucible-steel, and produces special grades of hard or high-carbon steel for tools, hammer- and piston-rods, etc. Fig. 14 shows the interior of these works.

The Damascus Steel Co., of Pittsburgh, makes a specialty of steel containing large percentages of nickel for cutlery, bicycle-spokes and chains, and extra-hard tool-steel for lathes and planers. It produces also a self-hardening nickel-alloy resembling Mushet or tungsten steel.

The Pope Tube Co., of Hartford, Conn., manufactures nickel-steel tubing for bicycles, and possesses every appliance for the production of nickel-steel tubing for light boilers, gas-engines, and all kinds of motors.

The Trenton Iron Co., of Trenton, N. J., manufactures "Tico" wire, a high-grade nickel-steel wire, for heaters and resistance-wire; and

The John A. Roebling Sons Co., of the same place, manufactures and sells a similar product, under the name of the "Climax Resistance Wire."

VIII.—BIBLIOGRAPHY.

J. Riley, " Alloys of Nickel and Steel," *Jour. Iron and Steel Inst.*, May, 1889.

Henri Schneider, *U. S. Patents*, Nos. 415,654 and 415,655.

J. Hopkinson, " Magnetic Properties of Alloys of Nickel and Iron," *Proc. Royal Soc.*, vol. xlvii., p. 23; vol. xlvii., p. 138; vol. xlviii., p. 1.

Sir Henry Bessemer, " Nickel-Steel," *Iron Age*, Feb. 27, 1890.

The Year's Naval Progress, Washington, 1891, p. 303.

M. C. Wallrand, " Rolling Nickel-Steel," *L'Ancre de St. Dizier*, and *Jour. Iron and Steel Inst.*, 1891, p. 412.

Nickel-Steel, *Comptes Rendus*, vol. cxiii. (1891), p. 33.

Garnier, *Engineering*, 1891, p. 763.

Harrington, *Iron*, 1892, p. 360.

F. Lynwood Garrison, " Development of American Armor-Plate," *Jour. Franklin Inst.*, May, June, July, 1892.

Dr. H. Wedding, " Nickeleisen Legirungen," *Verhandlungen des Vereins zur Beförderung des Gewerbfleisses*, Jan., 1892.

R. B. Dashiels, " Development of Nickel-Steel Armor-Plate," *Engineering Magazine*, Sept., 1893.

R. W. Davenport, *Trans. Naval and Marine Eng'rs.*, vol. i., 1893.

W. L. Austin, "Nickel," *Proc. Colorado Scientific Soc.*, Dec. 4, 1893.

Secretary U. S. Navy's Report, Nov. 18, 1893.

Cholat und Harmet, "Nickel-Stahl," *Oesterr. Zeit. für Berg- und Hüttenw.*, 1894.

The Steamship, June, 1894.

Capt. W. T. Sampson, "Present Status of Face-Hardened Armor-Plate."
Trans. Soc. Naval Arch. and Marine Engrs., 1894.

Iron Age, Dec. 6, 1894.

"On Nickel-Steel" (A Digest of M. Moulan's Experiments), *Industries and Iron*, Nov. 2, Nov. 23, 1894.

"British Armor and Ordnance," *The Engineer*, March 23, 1894.

Commander J. G. Eaton, "The Advantages of Nickel-Steel over Ordinary Steel," published by The Canadian Copper Co., Cleveland, O.

F. L. Sperry, "Nickel and Nickel-Steel," *Trans. A. I. M. E.*, March, 1895.

H. A. Wiggin, "Nickel-Steel," *Jour. Iron and Steel Inst.*, Aug., 1895.

M. Moulan, on "Ferro-Nickel Alloys." See *Jour. Iron and Steel Inst.*, 1894, ii., p. 457; *Stahl und Eisen*, vii. (1895), p. 346; *Rev. Univ. des Mines*, xxvii., p. 152.

"Alloys of Nickel, Iron and Steel," *Iron*, vol. xi., p. 162.

Lieutenant Jaques, "Nickel-Steel," *Engineering Review*, Feb. 20, 1895.

"Mineral Industry," 1895, p. 440.

"Corrosion of Nickel-Steels," *Report of Dept. City Works*, Brooklyn, N. Y., 1895, pp. 101–104.

H. H. Campbell, "Nickel-Steel Compared with Open-Hearth Steel," *Trans. Am. Soc. Civ. Engrs.*, Oct., 1895.

"Cast Nickel-Steel for Propeller-Blades," *Zeitsch. der Verein der Ing.*, 1895, p. 352.

"Nickel-Steel for Cycle-Making," *The Ironmonger*, Sept. 14, 1895.

Otto Vogel, "Ueber Darstellung, Eigenschaft und Verwendung von Nickel-Stahl," *Stahl und Eisen*, Aug. 1, 1895.

Lieut. A. A. Akerman, "Face-Hardened Armor," *Proc. U. S. Naval Inst.*, vol. xxi., No. 1.

Dewar and Fleming, "Changes Produced in Magnetized Iron and Steel by Cooling," *Proc. Royal Soc.*, 1896.

Wm. Beardmore, "Nickel-Steel," *Trans. Inst. Engrs. and Shipbuilders of Scotland*, March 24, 1896. *Industries and Iron*, May, 1896.

Ainsworth, "Magnetic Properties of Nickel-Steels," *Proc. Royal Soc.*, vol. lxii., p. 210–223.

H. W. Raymond, "Nickel-Steel in Metallurgy, Mechanics and Armor," *Eng. Mag.*, Feb., 1897.

Wedding and Rudeloff, "Bericht des Sonderausschusses fur Eisenlegirungen," *Verhandlung des Vereins zur Beförderung des Gewerbfleisses*, 1892, Heft I., II.; 1893, Heft I.; 1894, Heft III.; 1896, Heft II. and VIII.; 1897, Heft VI. and VII.; 1898, Heft I., VI. and VII.

R. W. Davenport, "Steel for Marine Forgings," *Cassier's Magazine*, Aug., 1897.

H. K. Landis, "Nickel-Steel," *Scientific American*, Jan. 9, 1897.

H. F. J. Porter, "Fatigue of Metal," *Jour. Frank. Inst.*, Dec., 1897.

Ch. Ed. Guillaume, "Récherches sur les aciers au Nickel," *Comptes Rendus*, March 11, 1898; *American Manufacturer*, Oct. 15, 1897; published separately by Gauthier Villars et Fils, Paris, 1898.

"The Manufacture of Bicycle-Tubing," *Iron Age*, Jan. 7–14; March 4, 1897.

John Harvard Biles, "Ship-Building Materials and Workmanship," *Inst. Civil Eng.*, 1897. *Engineering*, July 2, 1897.

Maunsel White, "Nickel-Steel Rivets," *Jour. Amer. Soc. Naval Engrs.*, Nov., 1898.

R. A. Hadfield, "Alloys of Nickel and Iron," *Tr. Inst. Civil Engrs.*, London, vol. cxxxviii., Part IV., Session 1898–1899.

H. F. J. Porter, Discussion on Nickel-Steel, *Trans. Am. Ry. Mast. Mech. Assn.*, June, 1899.

www.ingramcontent.com/pod-product-compliance
Lightning Source LLC
Chambersburg PA
CBHW020336090426
42735CB00009B/1556